MATHEMATICAL LOGIC

Notes on mathematics and its applications

General editors: Jacob T. Schwartz, *Courant Institute of Mathematical Sciences*, and Maurice Lévy, *Université de Paris*

E. Artin ALGEBRAIC NUMBERS AND ALGEBRAIC FUNCTIONS
R. P. Boas COLLECTED WORKS OF HIDEHIKO YAMABE
R. A. Bonic LINEAR FUNCTIONAL ANALYSIS
R. B. Burckel WEAKLY ALMOST PERIODIC FUNCTIONS ON SEMIGROUPS
M. Davis A FIRST COURSE IN FUNCTIONAL ANALYSIS
M. Davis LECTURES ON MODERN MATHEMATICS
J. Eells, Jr. SINGULARITIES OF SMOOTH MAPS
K. O. Friedrichs ADVANCED ORDINARY DIFFERENTIAL EQUATIONS
K. O. Friedrichs SPECIAL TOPICS IN FLUID DYNAMICS
A. Guichardet SPECIAL TOPICS IN TOPOLOGICAL ALGEBRAS
M. Hausner and J. T. Schwartz LIE GROUPS; LIE ALGEBRAS
P. Hilton HOMOTOPY THEORY AND DUALITY
S. Y. Husseini THE TOPOLOGY OF CLASSICAL GROUPS AND RELATED TOPICS
F. John LECTURES ON ADVANCED NUMERICAL ANALYSIS
A. M. Krall STABILITY TECHNIQUES FOR CONTINUOUS LINEAR SYSTEMS
R. L. Kruse and D. T. Price NILPOTENT RINGS
P. Lelong PLURISUBHARMONIC FUNCTIONS AND POSITIVE DIFFERENTIAL FORMS
H. Mullish AN INTRODUCTION TO COMPUTER PROGRAMMING
D. Ponasse MATHEMATICAL LOGIC
F. Rellich PERTURBATION THEORY OF EIGENVALUE PROBLEMS
J. T. Schwartz DIFFERENTIAL GEOMETRY AND TOPOLOGY
J. T. Schwartz NONLINEAR FUNCTIONAL ANALYSIS
J. T. Schwartz W-* ALGEBRAS
G. Sorani AN INTRODUCTION TO REAL AND COMPLEX MANIFOLDS
J. L. Soulé LINEAR OPERATORS IN HILBERT SPACE
J. J. Stoker NONLINEAR ELASTICITY
F. Trèves LINEAR PARTIAL DIFFERENTIAL EQUATIONS
F. Trèves LINEAR PARTIAL DIFFERENTIAL EQUATIONS WITH CONSTANT COEFFICIENTS

Additional volumes in preparation

Mathematical logic

DANIEL PONASSE

Professeur à la Faculté des Sciences de Lyon

GORDON AND BREACH
SCIENCE PUBLISHERS

New York · London · Paris

Gordon and Breach, Science Publishers, Inc.
440 Park Avenue South
New York, N.Y. 10016
First published in French as Logique Mathématique by OCDL, Paris

Editorial office for the United Kingdom
Gordon and Breach, Science Publishers Ltd.
42 William IV Street
London W.C. 2

Editorial office for France
Gordon & Breach
7–9 rue Emile Dubois
Paris 14e

Library of Congress catalog card number 72–136738. ISBN 0 677 30390 4 (cloth); 0 677 30395 5 (paper). Printed in east Germany.

Introduction

THIS BOOK is an introduction, assuming no previous acquaintance with logic. The mathematical prerequisites—in particular, the elements of set theory, algebra and topology—are limited to those covered in the *licence** course at French universities; the more specific concepts—Boolean rings, Boolean spaces—are not assumed known, and they are studied here in a simple but detailed manner.

The aim of the book is to present the two fundamental logical calculi:

The propositional calculus L';

the first-order predicate calculus L''.

These will be studied from four principal approaches, corresponding to different mathematical techniques:

(1) The syntactic, i.e., purely formal approach, which regards logic as a kind of game with elements and rules.

(2) The semantic approach, i.e., the problem of interpretation.

(3) The set-theoretic approach, which essentially involves the study of deductive systems.

(4) The algebraic approach and the topological approach, which enable one to utilize powerful and rapid methods.

The two essential problems of logic—consistency and completeness—have been carefully brought to the fore. The former is tackled very quickly; this is important, since it concerns the noncontradictoriness of the calculi under consideration, and its solution involves only elementary methods. The second problem, which is far more delicate (particularly with regard to the predicate calculus), is treated only at the end of the book, by algebraic and topological methods.

A distinctive feature of the book is the simultaneous treatment of the calculi L' and L'' (whenever this is possible), which has also dictated our choice of notation and terminology. The essential differences between these two calculi come to light in the investigation of complete deductive systems and valuations.

---

* *Translator's Note* Approximately equivalent to a Master's degree in the U.S.A.

I wish to thank my students at the Lyon Faculty of Sciences, especially Mess. Bourtot and Cusin, who were of great assistance in editing some of the chapters.

D. PONASSE

Translator's preface

OUR TRANSLATION OF Professor Ponasse's *Logique mathématique* follows the original in all particulars, except where we have changed the terminology to conform with that accepted in standard English texts on mathematical logic. In one case (the term *modèle*, p. 107), we have retained the author's term, using "model", though it is used in an entirely different sense in most English texts. We have also retained the author's use of formulas such as A, x and A, B to denote the unions $A \cup \{x\}$ and $A \cup B$, respectively.

To facilitate cross-references, we have made slight changes in the layout of the book, enumerating Propositions, Theorems, etc.

Contents

CHAPTER I

Syntactic approach: the calculi L' and L''

1 The propositional calculi L′ and L″

Symbolism

OUR ALPHABET, i.e., the set of symbols which we shall use, consists of the following elements:

1 Propositional variables The symbols u, v, w, ... (with or without subscripts or primes). The set of these symbols, assumed infinite, will be denoted by $\mathfrak{a}$.

2 Logical symbols (or connectives) The following five symbols:
$\neg$: negation, read "not"
$\wedge$: conjunction, read "and"
$\vee$: disjunction, read "or"
$\rightarrow$: conditional, read "implies"
$\leftrightarrow$: biconditional, read "is equivalent to".

For reasons of convenience which will become clear later we shall use the letters k, k', ... to represent any of the four connectives $\wedge$, $\vee$, $\rightarrow$, $\leftrightarrow$.

3 Punctuation symbols The following two symbols:
(: left parenthesis
): right parenthesis.

Note that these parentheses are formal symbols, not to be confused with ordinary parentheses.

This, then, is the sum total of our material, and in all rigor we should be able to describe the entire theory using this alphabet alone; however, in so doing one soon arrives at very long expressions, which are not easy to read. Moreover, it will often be necessary to represent the entities being constructed without writing them down explicitly. We shall therefore supplement the proper symbolism just defined by a second alphabet, exterior to our system, which we shall call a *metasymbolism.*

This metasymbolism will be infinite, and we shall introduce it as the need arises; in fact, we have already used metasymbols—the letters $\mathfrak{a}$, k, k'. We shall often employ the metasymbol $=$, for example, in the following sense:

$$A = (u) \wedge (v),$$

which means that the expression (see below) $(u) \wedge (v)$ will be represented by the metasymbol A.

Expressions Having an alphabet at our disposal, we shall now utilize it to construct words, which will be called expressions.

We define an expression to be any finite sequence of symbols written after the fashion of a word in the English language, i.e., from left to right in horizontal juxtaposition.

Example $)(u \rightarrow uv \neg$

We shall call each appearance of a symbol in an expression an *occurrence* of the symbol; occurrences may be enumerated and indexed (from left to right).

In the preceding example, there are two occurrences of u (one in the third position, the other in the fifth), and each of the other symbols has a single occurrence.

Sentences

Having defined expressions, i.e., words, we shall now divide them into two categories: words which have a "meaning" (in our theory), and words which have none; the former will be called sentences. Sentences will be defined constructively; to this end we first introduce the following concept.

Formation sequences A formation sequence is defined to be any finite sequence of expressions $A_1, A_2, \ldots, A_n$ (each metasymbol A_i represents an expression) such that each expression A_i satisfies (at least) one of the following conditions:

(1) A_i is a propositional variable;
(2) there exists $j < i$ such that $A_i = \neg(A_j)$;
(3) there exist $j < i$ and $h < i$ such that $A_i = (A_j)\, k\, (A_h)$.

To be precise, any such sequence will be called a formation sequence of length n for its last term A_n.

A *sentence* is defined to be an expression which has at least one formation sequence.

In particular, every propositional variable is a sentence, which is called an *atomic sentence.*

The following assertions are immediate.

If $A_1, \ldots, A_n$ is a formation sequence for A_n, then $A_1, \ldots, A_p$ is a formation sequence for A_p, for any $p \leqslant n$.

Thus every expression occurring in a formation sequence is a sentence.

If A is a sentence, then $\neg(A)$ is a sentence (add the latter expression to a formation sequence for A).

If A and B are sentences, then $(A)\,k\,(B)$ is a sentence (add the latter expression to the sequence obtained by writing a formation sequence for B after a formation sequence for A).

A sentence has a formation sequence of length 1 if and only if it is an atomic sentence.

For every sentence one can define a *minimal formation sequence* as a formation sequence of minimal length (the sentences in such a sequence are necessarily different).

We shall denote the set of sentences by E. The metasymbolic notation $A \in E$ thus means: A is a sentence.

Example of a formation sequence: u, v, $\neg(u)$, $(\neg(u)) \rightarrow (v)$.

According to the above definitions, every sentence is an expression of one of the following forms:

u where u is a propositional variable;
$\neg(B)$ where B is a sentence;
$(B)k(C)$ where B and C are sentences.

The problem arises as to whether this form is unique for a given sentence.

If A is a sentence, the following assertions are immediate:

(1) If A has the form u, it does not have another form v, $\neg(B)$, or $(B)k(C)$, since these forms have a different first symbol.

(2) If A has the form $\neg(B)$, it does not have the form u or $(B)k(C)$; if it has another form $\neg(C)$, then $B) = C)$, whence $B = C$.

(3) If A has the form $(B)k(C)$, it does not have the form u or $\neg(B)$.

However, in the latter case it is much less evident that the sentence cannot be expressed as $(B')k'(C')$ in a different way. In other words, does the distribution of the parentheses yield a unique decomposition of A into subsentences B and C?

To solve this problem, we introduce a few definitions and propositions.

Weight of a symbol: $p(\neg) = 1$,

$$p(\wedge) = p(\vee) = p(\rightarrow) = p(\leftrightarrow) = 2,$$

$$p(s) = 0 \quad \text{for any other symbol } s.$$

Weight of an expression α: $p(\alpha) =$ the sum of the weights of the symbols (different or not) comprising α.

Length of an expression α: $L(\alpha)$ = the number of symbols (different or not) comprising α.

Order of an expression α: $o(\alpha)$ = the number of connectives (different or not) involved in α.

$g(\alpha)$ will denote the number of left parentheses in the expression α.

$d(\alpha)$ will denote the number of right parentheses in the expression α.

If $\alpha = s_1 \cdots s_n$ (where the s_i are symbols), $n \geqslant 2$, a proper initial segment α' of the expression α is any expression $\alpha' = s_1 \cdots s_p$, where $p < n$.

1 Proposition *For any sentence A,*

$$g(A) = d(A) = p(A),$$

$$L(A) = 3p(A) + 1.$$

Proof By induction on $L(A)$.

If $L(A) = 1$, then $A = u$, $g(A) = d(A) = p(A) = 0$.

Assume the assertion true for any sentence of length less than n and let A be a sentence of length n.

If A is $\neg(B)$, then, by the induction hypothesis:

$$g(A) = g(B) + 1,$$

$$d(A) = d(A) + 1,$$

$$p(A) = p(A) + 1,$$

$$L(A) = L(B) + 3 = 3p(B) + 4 = 3p(A) + 1.$$

If A is $(B)k(C)$, then, again by the induction hypothesis:

$$g(A) = g(B) + g(C) + 2,$$

$$d(A) = d(B) + d(C) + 2,$$

$$p(A) = p(B) + p(C) + 2,$$

$$L(A) = L(B) + L(C) + 5 = 3p(B) + 3p(C) + 7 = 3p(A) + 1.$$

Thus the assertion is true for every sentence.

2 Proposition *If A is a sentence such that* $L(A) \geqslant 2$, *then, for any proper initial segment* α *of A, either* $g(\alpha) > d(\alpha)$ *or* $g(\alpha) = d(\alpha) = p(\alpha) \pm 1$.

Proof By induction on $L(A)$. By Proposition 1, the minimum length of a sentence which is not of length 1 is 4; since the only connective that such a sentence can contain is $\neg$, it must be $\neg(u)$ where u is a propositional variable.

If $A = \neg(u)$, the assertion is true, since α must be one of the expressions $\neg$, $\neg($, or $\neg(u$.

Assume the assertion true for every sentence of length less than n, and let A be a sentence of length n.

If $A = \neg(B)$:

$$\alpha = \neg \qquad g(\alpha) = d(\alpha) = 0 = p(\alpha) - 1;$$

$$\alpha = \neg(\qquad g(\alpha) = 1 > d(\alpha) = 0;$$

$\alpha = \neg(\beta$ where β is a proper initial segment of B:

$$g(\alpha) = g(\beta) + 1 > d(\beta) = d(\alpha);$$

$$\alpha = \neg(B \qquad g(\alpha) = g(B) + 1 = d(B) + 1 = d(\alpha) + 1.$$

If A is of the form $(B)k(C)$:

$$\alpha = (\qquad g(\alpha) = 1 > d(\alpha) = 0;$$

$$\alpha = (\beta \qquad g(\alpha) = g(\beta) + 1 > g(\beta) = d(\alpha);$$

$$\alpha = (B \qquad g(\alpha) = g(B) + 1 = d(B) + 1 = d(\alpha) + 1;$$

$$\alpha = (B) \qquad g(\alpha) = d(\alpha) = g(B) + 1 = p(B) + 1 = p(\alpha) + 1;$$

$$\alpha = (B)k \qquad g(\alpha) = d(\alpha) = g(B) + 1 = p(B) + 1 = p(\alpha) - 1;$$

$$\alpha = (B)k(\qquad g(\alpha) = g(B) + 2 = d(B) + 2 = d(\alpha) + 1;$$

$$\alpha = (B)k(\gamma \qquad g(\alpha) = g(B) + g(\gamma) + 2$$

$$d(\alpha) = d(B) + d(\gamma) + 1 \quad \text{and} \quad d(\gamma) \leqslant g(\gamma);$$

$$\alpha = (B)k(C \qquad g(\alpha) = g(B) + g(C) + 2$$

$$= d(B) + d(C) + 2 = d(\alpha) + 1.$$

Thus the assertion is true for all sentences.

Remark Every proper final segment α' of a sentence A has the property

$$g(\alpha') < d(\alpha') \quad \text{or} \quad g(\alpha') = d(\alpha') = p(\alpha') \pm 1,$$

for A may be expressed as $\alpha\alpha'$, where α is a proper initial segment, so that

$$g(\alpha') = g(A) - g(\alpha),$$

$$d(\alpha') = g(A) - d(\alpha), \quad \text{thus} \quad g(\alpha') < d(\alpha') \quad \text{since} \quad g(\alpha) > d(\alpha);$$

$$p(\alpha') = g(A) - p(\alpha), \quad \text{thus} \quad g(\alpha') = d(\alpha') = p(\alpha') \pm 1,$$

$$\text{since} \quad g(\alpha) = d(\alpha) = p(\alpha) \pm 1.$$

COROLLARY *No proper initial or final segment of a sentence is a sentence.*

Now suppose that a sentence A can be expressed in two different forms:

$$A = (B)k(C) = (B')k'(C').$$

If $L(B) = L(B')$, then $B = B'$, and thus $k = k'$ and $C = C'$; if, say, $L(B') < L(B)$, then B' is a proper initial segment of B, contradicting the fact that B' is a sentence.

Remarks The two propositions established above are necessary conditions for an expression to be a sentence; however, they are not sufficient, as may be seen by considering the expression $\neg u()$, which is not a sentence.

A natural problem is to determine necessary and sufficient conditions which provide a purely intrinsic characterization of sentences. However, this would be rather difficult in the system of notation adopted here (and of limited interest for the sequel), though we may mention that it has been done for other systems, notably the parenthesis-free notation of Łukasiewicz.

To summarize, every sentence may be expressed in one and only one of the forms u, $\neg(B)$, $(B)k(C)$, where the propositional variable u, the sentences B and C, and the connective k are uniquely determined (independently of any formation sequence for A).

It is thus possible to define the *dominant logical symbol* of a sentence A:

An atomic sentence has no dominant symbol;

if $A = \neg(B)$, then $\neg$ is its dominant logical symbol;

if $A = (B)k(C)$, then k is its dominant logical symbol.

Similarly, one can define a *proper subsentence* of a sentence A:

An atomic sentence has no proper subsentence;

if $A = \neg(B)$, then B is a proper subsentence;

if $A = (B)k(C)$, then B and C are proper subsentences;

any proper subsentence of a proper subsentence of A is a proper subsentence of A; no other expression is a proper subsentence of A.

A *subsentence* of a sentence A is any sentence which is either A itself or a proper subsentence of A. The set of subsentences of A will be denoted by S_A.

Remark 1 As an exercise the reader may show that every minimal formation sequence of a sentence A contains all subsentences of A, each occurring exactly once, and nothing else. Consequently, two minimal formation sequences for the same sentence A may only differ in the order of their terms.

Remark 2 Let A be a sentence of length at least 2,

$$A = s_1 s_2 \cdots s_n.$$

By the proof of Proposition 2, there exists a unique proper initial segment α such that

$$g(\alpha) = d(\alpha) = p(\alpha) - 1;$$

if $\alpha = s_1 \cdots s_m$, then s_m is the dominant logical symbol of A.

Simplified notation for sentences To facilitate the writing of sentences, we shall now adopt a number of conventions regarding the omission of parentheses:

Omission of parentheses around a propositional variable:

Example $u \to v$ instead of $(u) \to (v)$.

Omission of parentheses in repeated negations:

Example $\neg\neg\neg(A)$ instead of $\neg(\neg(\neg(A)))$.

The connectives $\wedge$ $\vee$ $\to$ $\leftrightarrow$ will be considered to "rank above" $\neg$:

Example $\neg u \to v$ instead of $(\neg u) \to v$.

The connectives $\to$ and $\leftrightarrow$ will be considered to "rank above" $\wedge$ and $\vee$:

Example $u \wedge v \to w$ instead of $(u \wedge v) \to w$.

Of course, certain parentheses will still be needed, for example, in the formula $u \wedge (v \to w)$.

This notion of the "rank" of symbols is entirely exterior to our system; it belongs to the metasymbolism, and results in sentences expressed in a metanotation for which Propositions 1 and 2 are no longer valid.

Substitution Let A and B be two sentences and u a propositional variable. We define the result of substituting B for u in A as the expression obtained from A by replacing each occurrence of u in A by B. Here we are assuming that A and B are expressed in full notation (i.e., without omission of parentheses). The resulting expression will be denoted by $(B/u)\,A$.

Reasoning by induction on $L(A)$, we prove that $(B/u)\,A$ is also a sentence.

If $L(A) = 1$, then A is a propositional variable:

$$\text{If} \quad A = u, \quad \text{then} \quad (B/u)\,A = B;$$

$$\text{if} \quad A = v, \quad \text{then} \quad (B/u)\,A = v.$$

Assume the assertion true for every sentence of length than n, and let A be a sentence of length n.

If $A = \neg(A')$, then $(B/u)\,A = \neg((B/u)\,A')$ and $(B/u)\,A'$ is a sentence.

If $A = (A')k(A'')$, then $(B/u)\,A = ((B/u)\,A')k((B/u)\,A'')$, and $(B/u)\,A'$ and $(B/u)\,A''$ are sentences.

Remark In practice, substitutions are of course performed on sentences in simplified notation.

Example $(u \wedge v/w)\,(w \to w \vee u) = u \wedge v\,(u \wedge v) \vee u.$

Axioms, provable sentences

Continuing our exposition of the rules of the game, we shall now stipulate that certain sentences are "true" (which reduces to defining the meaning of the word "true"). Consider three different propositional variables u^0, v^0, w^0, chosen once and for all (*a priori* this choice conditions the following arguments, but in fact we shall soon see that it is immaterial). The following nine expressions, easily seen to be sentences in simplified notation, are called *axioms*:

$$\begin{array}{ll}
(1) & u^0 \to (v^0 \to u^0) \\
(2) & (u^0 \to (u^0 \to v^0)) \to (u^0 \to v^0) \\
(3) & (u^0 \to v^0) \to ((v^0 \to w^0) \to (u^0 \to w^0)) \\
(4) & (u^0 \leftrightarrow v^0) \to (u^0 \to v^0) \\
(5) & (u^0 \leftrightarrow v^0) \to (v^0 \to u^0) \\
(6) & (u^0 \to v^0) \to ((v^0 \to u^0) \to (u^0 \leftrightarrow v^0)) \\
(7) & (\neg v^0 \to \neg u^0) \to (u^0 \to v^0) \\
(8) & u^0 \vee v^0 \leftrightarrow (\neg u^0 \to v^0) \\
(9) & u^0 \wedge v^0 \leftrightarrow (\neg u^0 \vee v^0).
\end{array}$$

We now define a *formal proof* to be any finite sequence of sentences $A_1, \ldots, A_n$, each of which satisfies (at least) one of the following three conditions:

(1) A_i is an axiom;

(2) there exist $j < i$, a sentence B, and a propositional variable u such that $A_i = (B/u)\,A_j$;

(3) there exist $j < i$ and $h < i$ such that $A_h = A_j \to A_i$.

Such a sequence is said to be a formal proof, of length n, for its last sentence A_n.

We define a *provable sentence* to be a sentence which has at least one formal proof. The notation $\vdash A$ means that A is a provable sentence.

If we denote the set of all provable sentences by T, an alternative notation is $A \in T$.

Remark 1 If $A_1, \ldots, A_n$ is a formal proof of A_n, then $A_1, \ldots, A_p$ is a formal proof of A_p for any $p \leqslant n$.

Remark 2 In particular, every sentence occurring in a formal proof is a provable sentence.

Remark 3 We have the following two rules (called rules of inference):
If $\vdash A$ and $\vdash A \to B$, then $\vdash B$ (rule of *modus ponens*) (a formal proof of the latter sentence consists of a proof of A and a proof of $A \to B$, followed by B).

If $\vdash A$, then $\vdash (B/u)\, A$ for any sentence B and propositional variable u (rule of *substitution*) (add the formula $(B/u)\, A$ at the end of a proof of A).

Example of a formal proof

(1)	$u^0 \to (v^0 \to u^0)$	Axiom 1
(2)	$u^0 \to (u^0 \to u^0)$	(u^0/v^0) 1
(3)	$(u^0 \to (u^0 \to v^0)) \to (u^0 \to v^0)$	Axiom 2
(4)	$(u^0 \to (u^0 \to u^0)) \to (u^0 \to u^0)$	(u^0/v^0) 3
(5)	$u^0 \to u^0$	modus ponens (2, 4)
(6)	$(u^0 \leftrightarrow v^0) \to (v^0 \to u^0)$	Axiom 5
(7)	$(w^0 \leftrightarrow v^0) \to (v^0 \to w^0)$	(w^0/u^0) 6
(8)	$(w^0 \leftrightarrow s^0) \to (s^0 \to w^0)$	(s^0/v^0) 7, $s^0 \neq w^0$
(9)	$(u^0 \vee v^0 \leftrightarrow s^0) \to (s^0 \to u^0 \vee v^0)$	$(u^0 \vee v^0/w^0)$ 8
(10)	$(u^0 \vee v^0 \leftrightarrow (\neg u^0 \to v^0)) \to ((\neg u^0 \to v^0) \to u^0 \vee v^0)$	$(\neg u^0 \to v^0/s^0)$ 9
(11)	$u^0 \vee v^0 \leftrightarrow (\neg u^0 \to v^0)$	Axiom 8
(12)	$(\neg u^0 \to v^0) \to u^0 \vee v^0$	modus ponens (10, 11)
(13)	$(\neg u^0 \to \neg u^0) \to u^0 \vee \neg u^0$	$(\neg u^0/v^0)$ 12
(14)	$\neg u^0 \to \neg u^0$	$(\neg u^0/u^0)$ 5
(15)	$u^0 \vee \neg u^0$	modus ponens (13, 14)

In particular: $\vdash A \to A$

$$\vdash A \vee \neg A$$

(principle of the excluded middle) for any formula A (substitute A for u^0 in 5 and 15).

It is clear that this procedure rapidly becomes very tedious. However, we shall see that it may be considerably abbreviated. To this end we first establish the following result:

3 Proposition *If A is a provable sentence, it possesses a formal proof (not necessarily unique) which begins with a sequence* (α) *of axioms, followed by a sequence* (σ) *of sentences derived from previous sentences by substitution, and finally a sequence* (μ) *of sentences each derived from two previous sentences by modus ponens.*

A proof of this type will be represented by the symbol

$$\Sigma^* = (\alpha)\,(\sigma)\,(\mu)$$

(the sequences (σ) or (μ) may be empty).

Proof Any formal proof of A includes a certain number of axioms (at least one). Bringing these axioms to the beginning of the sequence (in the same relative order) we obtain a new sequence:

$$\Sigma_1 = (\alpha)\Sigma_1'$$

where Σ_1' contains no axioms.

The sequence Σ_1 is again a proof, since the sentences preceding any term of Σ_1 which is not an axiom include those preceding it in Σ_1.

If $A \in (\alpha)$ the assertion is proved. Otherwise A is the last term of Σ_1'.

Every formula B in Σ_1' is obtained either by substitution of a propositional variable in a previous sentence or by modus ponens from two previous sentences (if any sentence falls into both these categories we choose one of them arbitrarily).

We now classify the sentences in Σ_1' as follows:

(1) Simple substitution instances of axioms, i.e., sentences obtained by substitution of a propositional variable in an axiom; let (b_1) denote the subsequence consisting of these sentences.

(2) Double substitution instances of axioms, i.e., sentences obtained by substitution in a sentence of (b_1); let (b_2) denote the subsequence consisting of these sentences.

We continue this classification in an obvious manner.

Let (σ') be the subsequence formed by juxtaposing the subsequences (b_i) (i.e., all multiple substitution instances of axioms) in order. Bringing all these sentences to the beginning of Σ_1', we obtain a sequence

$$\Sigma_2 = (\alpha)\,(\sigma')\,\Sigma_2'.$$

The sequence Σ_2 is again a formal proof, and we assume that Σ_2' is not empty (A is then its last term).

Let B_1 be the first sentence of Σ_2' (which is obtained by modus ponens), and let $\mathscr{B}_1$ be the subsequence consisting of B_1 followed by all its multiple substitution instances.

Let B_2 be the first sentence not yet classified (which is again obtained by modus ponens), and let $\mathscr{B}_2$ be the subsequence consisting of B_2 followed by all its multiple substitution instances. Continuing this construction in the same way, we obtain a formal proof

$$\Sigma_3 = (\alpha)\,(\sigma')\,\mathscr{B}_1 \cdots \mathscr{B}_s,$$

where each subsequence $\mathscr{B}_i$ begins with a sentence obtained by modus ponens followed (perhaps) by multiple substitution instances of that sentence.

Now consider the last subsequence $\mathscr{B}_s$, and let B_s be its first sentence, which is obtained in the following manner:

$$\ldots, C, \ldots, C \to B_s, \ldots, B_s.$$

Suppose that B_s is followed by a sentence D, which can only be a sentence of the form $(E/u)\,B$. We can then consider D as obtained by modus ponens, by inserting two additional sentences:

$$\ldots, C, (E/u)\,C, \ldots, C \to B_s, (E/u)\,C \to (E/u)\,B_s, \ldots, B_s, (E/u)\,B_s,$$

and the two added sentences are substitution instances, which are inserted in the subsequences preceding B_s. When this has been done for all simple substitution instances of B_s, the double substitution instances become simple substitution instances of sentences obtained by modus ponens, to which we can now apply the same procedure, and so on.

By successive application of this procedure to the subsequences preceding $\mathscr{B}_s$, we finally obtain a formal proof of the desired type.

Remark The above procedure is finite, as is easily verified: Let $n + 1$ be the maximum number of sentences in each of the original subsequences $\mathscr{B}_i$. The operations applied to the sentences in $\mathscr{B}_i$ result in the addition of at most $2n$ sentences to the preceding subsequences, then $6n$ sentences, ..., and finally the number of added sentences is at most $(3^s - 1)\,n$.

We now introduce the following definition:

Axiom schemata Let $\bar{A}, \bar{B}, \bar{C}$ be three different "letters"; then the following nine expressions are axiom schemata:

S1) $\bar{A} \to (\bar{B} \to \bar{A})$

S2) $(\bar{A} \to (\bar{A} \to \bar{B})) \to (\bar{A} \to \bar{B})$

S3) $(\bar{A} \to \bar{B}) \to ((\bar{B} \to \bar{C}) \to (\bar{A} \to \bar{C}))$

S4) $(\bar{A} \leftrightarrow \bar{B}) \to (\bar{A} \to \bar{B})$
S5) $(\bar{A} \leftrightarrow \bar{B}) \to (\bar{B} \to \bar{A})$
S6) $(\bar{A} \to \bar{B}) \to ((\bar{B} \to \bar{A}) \to (\bar{A} \leftrightarrow \bar{B}))$
S7) $(\neg\bar{B} \to \neg\bar{A}) \to (\bar{A} \to \bar{B})$
S8) $\bar{A} \vee \bar{B} \leftrightarrow (\neg\bar{A} \to \bar{B})$
S9) $\bar{A} \wedge \bar{B} \leftrightarrow \neg(\neg\bar{A} \vee \neg\bar{B})$.

An *instance of an axiom schema* is any expression obtained from an axiom schema when the letters are replaced by sentences.

The ith axiom schema will be denoted by

$$\langle \bar{A}, \bar{B}, \bar{C} \rangle_i.$$

4 PROPOSITION *Every instance of an axiom schema is a provable sentence.*

Proof Consider an instance of the axiom schema Si): $\langle A, B, C \rangle$; then the similarly numbered axiom $\langle u^0, v^0, w^0 \rangle$ is an instance of the same axiom schema. Let s^0 be a propositional variable different from u^0 and w^0 and not occurring in A, t^0 a propositional variable different from u^0 and s^0 and occurring neither in A nor in B. Then

$$(s^0/v^0)\,\langle u^0, v^0, w^0 \rangle = \langle u^0, s^0, w^0 \rangle$$
$$(t^0/w^0)\,\langle u^0, s^0, w^0 \rangle = \langle u^0, s^0, t^0 \rangle$$
$$(A/u^0)\,\langle u^0, s^0, t^0 \rangle = \langle A, s^0, t^0 \rangle$$
$$(B/s^0)\,\langle A, s^0, t^0 \rangle = \langle A, B, t^0 \rangle$$
$$(C/t^0)\,\langle A, B, t^0 \rangle = \langle A, B, C \rangle$$

This proves, first, that $\langle A, B, C \rangle$ is a provable sentence, and, second, that every instance of an axiom schema is a multiple substitution instance of an axiom.

Conversely, every axiom and every multiple substitution instance of an axiom is an instance of an axiom schema. This may be proved by induction on the number n of substitutions:

If $n = 0$ this is trivial.

If the assertion is assumed true for all substitution instances of order $n - 1$, and we consider a substitution instance of order n, the latter has the form $(B/u)\,A$, and by the induction hypothesis A is an instance of an axiom schema, say $A = \langle A', B', C' \rangle_i$. Then

$$(B/u)\,A = \langle (B/u)\,A', (B/u)\,B', (B/u)\,C' \rangle_i$$

which is clearly an instance of an axiom schema.

COROLLARY *In a formal proof of the type* $\Sigma^* = (\alpha)\,(\sigma)\,(\mu)$ *considered above, every sentence in* $(\alpha)\,(\sigma)$ *is an instance of an axiom schema.*

We can thus formulate an alternative definition of formal proof: A formal proof is a sequence $A_1, \ldots, A_n$ of sentences each of which satisfies at least one of the following two conditions:

(1) A_i is an instance of an axiom schema.

(2) A_i is derived by modus ponens from two previous sentences.

This is actually another type of formal proof; however, the propositions established above show that a sentence has a formal proof in the first sense if and only if it has one in the second sense. In the sequel, therefore, we shall only employ the second type of formal proof.

Example It is now much easier to prove the principle of the excluded middle:

(1)	$\neg A \to (\neg A \to \neg A)$	instance of S1
(2)	$(\neg A \to (\neg A \to \neg A)) \to (\neg A \to \neg A)$	instance of S2
(3)	$\neg A \to \neg A$	modus ponens (1, 2)
(4)	$(A \vee \neg A \leftrightarrow (\neg A \to \neg A)) \to ((\neg A \to \neg A) \to A \vee \neg A)$	instance of S5
(5)	$A \vee \neg A \leftrightarrow (\neg A \to \neg A)$	instance of S8
(6)	$(\neg A \to \neg A) \to A \vee \neg A$	modus ponens (4, 5)
(7)	$A \vee \neg A$	modus ponens (3, 6)

2 The predicate calculus L''

We shall now introduce another type of logical calculus; our description will be analogous to that of the propositional calculus. The new calculus is known as the restricted first-order predicate calculus without equality.

Symbolism

The alphabet is the union of the following sets:

1 Individual symbols or individuals an infinite set of elements denoted by $a, b, c, \ldots$ This set will be denoted by I.

2 Figurative symbols or variables an infinite set of elements denoted by $x, y, z, \ldots$ This set will be denoted by X.

3 Relation symbols or predicates a set of elements denoted by r_n^p, where p and n are natural numbers; the superscript p is called the *weight* of the predicate (its interpretation will be explained later), and the subscript n distinguishes between different predicates of the same weight. We shall assume that the alphabet contains at least one predicate of nonzero weight. The set of predicates is not assumed denumerable, the notation notwithstanding; neither is it assumed infinite.

4 *Logical symbols or connectives* The seven connectives are:

(a) the five connectives of the calculus L': $\neg \wedge \vee \rightarrow \leftrightarrow$; we shall again use the notation $k, k', \ldots$ for any of the four connectives $\wedge \vee \rightarrow \leftrightarrow$;

(b) the existential symbol $\exists$, read "there exists";

(c) the universal symbol $\forall$, read "for all"; we shall use the letters $Q, Q', \ldots$ to denote either of the last two connectives.

5 *Punctuation symbols* parentheses and square brackets: ( ) [ ].

Of course, apart from the alphabet proper we shall make extensive use of a metabolism, to be defined as the need arises.

Expressions An expression is any finite sequence of symbols, such as $) \exists\, r_2^7 a \rightarrow$. We can again speak of an occurrence of a symbol in an expression.

Substitution of symbols Given an expression α and two symbols σ and σ', we define the result of substituting σ' for σ in α as the expression α' obtained from α when σ is replaced by σ' at each of its occurrences. Notation:

$$\alpha' = (\sigma'/\sigma)\,\alpha.$$

Example $(x/a)\,(r_0^2 ab \rightarrow r_0^1 a) = r_0^2 xb \rightarrow r_0^1 x$.

We shall often use the following notation:

$\alpha\, \rangle \sigma_1, \ldots, \sigma_n \langle$ denotes an expression α which does not contain the symbols $\sigma_1, \ldots, \sigma_n$;

$\alpha\, \langle \sigma_1, \ldots, \sigma_n \rangle$ denotes an expression α which contains the symbols $\sigma_1, \ldots, \sigma_n$ (and possibly others).

Remark In general, substitutions will only be performed on individuals and variables.

Sentences and formulas

We define an *atomic sentence* to be any expression consisting of a predicate r_n^p of weight p followed by p individuals:

$$r_n^p a_1 \cdots a_p.$$

As in the propositional calculus, atomic sentences will be denoted by $u, v, w, \ldots$ and the set of all atomic sentences by $\mathfrak{a}$.

Quantifiers For any variable x:

the expression $\exists x$ will be called the existential quantifier of x;

the expression $\forall x$ will be called the universal quantifier of x.

In general, a quantifier is any expression which is the existential or universal quantifier of a variable: Qx.

A *formation sequence* is any finite sequence of expressions $A_1, \ldots, A_n$ each of which satisfies (at least) one of the following conditions:

(1) A_i is an atomic sentence;

(2) there exists $j < i$ such that $A_i = \neg(A_j)$;

(3) there exist $j < i$ and $h < i$ such that $A_i = (A_j)k(A_h)$;

(4) there exist $j < i$, an individual a, and a variable x such that

$$A_j \rangle x \langle \quad \text{and} \quad A_i = Qx\,[(x/a)\,A_j].$$

Such a sequence is said to be a formation sequence, of length n, for its last term A_n; a *sentence* is any expression which has a formation sequence.

Remark 1 Any expression in a formation sequence is a sentence.

Remark 2 If A is a sentence, so is $\neg(A)$.

Remark 3 If A and B are sentences, so is $(A)k(B)$.

Remark 4 If A is a sentence, so is $Qx\,[(x/a)\,A]$ for any x such that $A \rangle x \langle$ and any individual a.

Remark 5 We define the *order* of a sentence, denoted $o(A)$, as the number of occurrences of logical symbols in A.

Remark 6 The *basis* of a sentence A, denoted by I_A, is the (finite) set of individuals with at least one occurrence in A.

Example of a formation sequence:

$$r_0^2\,ab$$

$$\exists x\,[r_0^2\,xb]$$

$$(r_0^2\,ab) \to (\exists x\,[r_0^2\,xb])$$

$$\forall y\,[(r_0^2\,ay) \to (\exists x\,[r_0^2\,xy])].$$

We denote the set of sentences by E.

We shall adopt the same conventions for omission of parentheses as in the propositional calculus, and in addition omit the external parentheses in $(Qx\,[(x/a)\,A])$.

Example $\forall y\,[r_0^2\,ay \rightarrow \exists x\,[r_0^2\,xy]]$.

A *quantifiable formula* with respect to a variable x is any expression of the form $(x/a)\,A$ where A is a sentence such that $A\,\rangle x\langle$; we shall usually use the notation f^x for a quantifiable formula.

A *formulation* is any finite sequence of expressions $f_1, \ldots, f_n$ such that each expression f_i satisfies at least one of the following conditions:

(1) f_i is a quantifiable formula with respect to some variable;
(2) there exists $j < i$ such that $f_i = (x/a)\,f_j$ where $f_j\,\rangle x\langle$.

A *formula* is any expression which has a formulation. We define the order and basis of a formula as before.

Remark Every sentence is a formula, since $A = (x/a)\,A$ where $A\,\rangle a\langle$.

Examples of formulas:

$r_0^2\,xb$	(quantifiable formula with respect to x)
$r_0^2\,xy$	
$\exists x\,[r_0^2\,xy]$	(quantifiable formula with respect to y).

A variable x is said to be *bound* in a formula if the latter contains at least one occurrence of Qx; otherwise the variable is said to be *free*. A sentence contains no free variables.

Thus every sentence must have one of the following four forms:

u	(where $u \in \mathfrak{a}$),
$\neg(A)$, $(A)k(B)$	(where A and B are sentences),
$Qx\,[f^x]$	(where f^x is a quantifiable formula with respect to x).

As before, it can be proved that any given sentence is uniquely expressible in one of these forms.

Remark Our last assertion should be understood in the sense that the form of a sentence is unique *once all substitutions of variables and individuals have been performed*, since it may happen that

$$Qx\,[(x/a)\,B] = Qx\,[(x/b)\,C];$$

for example,

$$\exists x\,[r_0^2\,xb] = \exists x\,[(x/a)\,r_0^2\,ab] = \exists x\,[(x/c)\,r_0^2\,cb].$$

This shows that a sentence may have infinitely many formation sequences (even minimal ones) containing different sentences.

5 Proposition *If the expression α is a sentence [quantifiable formula] then, for any individuals a and b, the expression $(a/b)\,\alpha$ is also a sentence [quantifiable formula].*

Proof by induction on the order n of α.

If α is a sentence of order 0, i.e., an atomic sentence, it is immediate that $(a/b)\,\alpha$ is also an atomic sentence.

Let α be a quantifiable formula of order 0, i.e., $\alpha = (x/c)\,u$ where u is an atomic sentence. Then:

$$(a/b)\,\alpha = (a/b)\,(x/c)\,u.$$

If $b = c$ then $(a/b)\,\alpha = (a/b)\,(x/b)\,u = (x/b)\,u = \alpha$.

If $b \neq c$, let d be an individual different from a, b, c which does not occur in u; then

$$(x/c)\,u = (x/d)\,(d/c)\,u,$$

$$(a/b)\,\alpha = (a/b)\,(x/d)\,(d/c)\,u = (x/d)\,(a/b)\,(d/c)\,u.$$

Now u, therefore also $(d/c)\,u$ and $(a/b)\,(d/c)\,u$, are atomic sentences (not containing x), so that $(a/b)\,\alpha$ is indeed a quantifiable formula with respect to x.

Now assume the assertion true for all sentences and quantifiable formulas of order at most n, and let α be a sentence of order $n+1$.

If $\alpha = \neg(A)$, then $(a/b)\,\alpha = ((a/b)\,A)$, which is a sentence.

If $\alpha = (A)k(B)$, then $(a/b)\,\alpha = ((a/b)\,A)k((a/b)\,B)$, which is a sentence.

If $\alpha = Qx\,[f^x]$, then $(a/b)\,\alpha = Qx\,[(a/b)f^x]$, which is a sentence.

Thus the assertion is true for all sentences of order $n+1$.

Finally, if α is a quantifiable formula of order $n+1$, $\alpha = (x/c)\,A$, the reasoning is exactly the same as in the case of order 0 (replacing u by A).

This result may be generalized by considering what we call a *simultaneous substitution* of individuals, i.e., any mapping $s : I \hookrightarrow I$. For any expression α, we can then define the expression $s(\alpha)$ obtained from α by replacing all the individuals by their images under s.

6 Proposition *If A is a sentence, $s(A)$ is also a sentence.*

Proof Let $I_A = \{a_1, \ldots, a_n\}$, and set $b_1 = s(a_1), \ldots, b_n = s(a_n)$.

Choose individuals $\{c_1, \ldots, c_n\}$ which are different from each other, different from the a_i and the b_i, and have no occurrences in A.

Then $s(A) = (b_n/c_n)(b_{n-1}/c_{n-1}) \cdots (b_1/c_1)(c_n/a_n)(c_{n-1}/a_{n-1}) \cdots (c_1/a_1) A$.

Similarly, let f^x be a quantifiable formula with respect to x,

$$f^x = (x/a)\,A.$$

Let $I_A = \{a_1, \ldots, a_n\}$ and choose an individual b different from $s(a_1), \ldots, s(a_n)$ which does not occur in A. Now consider the simultaneous substitution: $s'(c) = s(c)$ if $c \neq b$, $s'(b) = b$. Then

$$s(f^x) = s'(f^x);$$

$$f^x = (x/b)(b/a)\,A = (x/b)\,A', \quad \text{where} \quad A' = (b/a)\,A.$$

Thus $s(f^x) = (x/b)\,s'(A')$, which is again a quantifiable formula with respect to x.

Remark If f^x is a quantifiable formula with respect to x, then for every individual a the expression $(a/x)f^x$ is a sentence, since $f^x = (x/b)\,A \rangle x \langle$.

Thus $(a/x)f^x = (a/b)(x/b)\,A = (a/b)\,A$.

Remark One can also prove that if f is a formula then so is $(a/b)f$.

One can again define the ***dominant logical symbol*** of a sentence A:

An atomic sentence has no dominant logical symbol;

if $A = \neg(B)$ then $\neg$ is its dominant logical symbol;

if $A = (B)k(C)$ then k is its dominant logical symbol;

if $A = Qx\,[f^x]$ then Q is its dominant logical symbol.

We now define a *proper subsentence* of a sentence A:

An atomic sentence has no proper subsentence;

if $A = \neg(B)$ then B is a proper subsentence of A;

if $A = (B)k(C)$ then B and C are proper subsentences of A;

if $A = Qx\,[f^x]$ then all sentences $(a/x)f^x$, where a is any individual, are proper subsentences of A;

any proper subsentence of a proper subsentence of A is a proper subsentence of A;

no other expression is a proper subsentence of A.

A subsentence of A is any proper subsentence of A or A itself. We denote the set of subsentences of A (which may be infinite) by S_A.

Remark In contrast to the situation in the calculus L', in the calculus L'' a subsentence need not occur in a formation sequence of the given sentence.

It may not even occur in any minimal formation sequence; for example, let

$$A = \exists x\,[r_0^2 ax],$$

then $r_0^2 aa$ is a subsentence which cannot occur in any minimal formation sequence.

More precisely: if $A = Qx\,[f^x]$, where $f^x = (x/b)\,B$, $B\,\rangle x\langle$, then $(b/x)f^x = B$, where $f^x\,\rangle b\langle$, is a subsentence which can occur in a minimal formation sequence.

Conversely, given a subsentence $(a/x)f^x$ where $f^x\,\rangle a\langle$, we have

$$f^x = (x/a)\,(a/x)f^x,$$

so that $(a/x)f^x$ can occur in a minimal formation sequence.

One might thus define "generating" subsentences of a sentence $Qx\,[f^x]$ as expressions $(a/x)f^x$ with $f^x\,\rangle a\langle$, but we shall not make this distinction in the sequel.

Axioms, provable sentences

The following eleven expressions will be called axiom schemata:

S1) $\bar{A} \to (\bar{B} \to \bar{A})$
S2) $(\bar{A} \to (\bar{A} \to \bar{B})) \to (\bar{A} \to \bar{B})$
S3) $(\bar{A} \to \bar{B}) \to ((\bar{B} \to \bar{C}) \to (\bar{A} \to \bar{C}))$
S4) $(\bar{A} \leftrightarrow \bar{B}) \to (\bar{A} \to \bar{B})$
S5) $(\bar{A} \leftrightarrow \bar{B}) \to (\bar{B} \to \bar{A})$
S6) $(\bar{A} \to \bar{B}) \to ((\bar{B} \to \bar{A}) \to (\bar{A} \leftrightarrow \bar{B}))$
S7) $(\neg \bar{B} \to \neg \bar{A}) \to (\bar{A} \to \bar{B})$
S8) $\bar{A} \vee \bar{B} \leftrightarrow (\neg \bar{A} \to \bar{B})$
S9) $\bar{A} \wedge \bar{B} \leftrightarrow \neg(\bar{A} \vee \neg \bar{B})$
S10) $(\bar{a}/\bar{x})f^{\bar{x}} \to \exists \bar{x}\,[f^{\bar{x}}]$
S11) $\forall \bar{x}\,[f^{\bar{x}}] \to (\bar{a}/\bar{x})f^{\bar{x}}$.

Note that the first nine axiom schemata are the same as those of the propositional calculus.

An *instance of an axiom schema* is any sentence obtained from an axiom schema by replacing the letters $\bar{A}$, $\bar{B}$, $\bar{C}$ by sentences, $\bar{a}$ by an individual, $\bar{x}$ by a variable x, and $f^{\bar{x}}$ by a quantifiable formula with respect to x (it may be verified that the result is indeed a sentence).

A *formal proof* is any finite sequence of sentences $A_1, \ldots, A_n$, each of which satisfies (at least) one of the following conditions:

(1) A_i is an instance of an axiom schema;

(2) there exist $j < i$ and $h < i$ such that $A_h = A_j \to A_i$;

(3) $A_i = \exists x\,[f^x] \to A$ and there exists $j < i$ such that $A_j = (a/x) f^x \to A$, where $A\,\rangle a \langle$ and $f^x\,\rangle a \langle$;

(4) $A_i = A \to \forall x\,[f^x]$ and there exists $j < i$ such that $A_j = A \to (a/x) f^x$ where $A\,\rangle a \langle$ and $f^x\,\rangle a \langle$.

Such a sequence will be called a formal proof of its last term A_n, of length n.

A *provable sentence* is any sentence which has a formal proof; we symbolize this fact by $\vdash A$. As before, T will denote the set of provable sentences.

Remark 1 If $\vdash A$ and $\vdash A \to B$ then $\vdash B$ (modus ponens).

Remark 2 If $\vdash (a/x) f^x\,\rangle a \langle \to A\,\rangle a \langle$ then $\vdash \exists x\,[f^x] \to A$.

Remark 3 If $\vdash A\,\rangle a \langle \to (a/x) f^x\,\rangle a \langle$ then $\vdash A \to \forall x\,[f^x]$.

Example of a formal proof

(1)	$r_0^2ab \to (r_0^2ab \to r_0^2ab)$	S1
(2)	$(r_0^2ab \to (r_0^2ab \to r_0^2ab)) \to (r_0^2ab \to r_0^2ab)$	S2
(3)	$r_0^2ab \to r_0^2ab$	m.p.
(4)	$r_0^2ab \to \exists x\,[r_0^2xb]$	S10
(5)	$\forall y\,[r_0^2ay] \to r_0^2ab$	S11
(6)	$(\forall y\,[r_0^2ay] \to r_0^2ab) \to ((r_0^2ab \to \exists x\,[r_0^2xb]) \to (\forall y\,[r_0^2ay] \to \exists x\,[r_0^2xb]))$	S3
(7)	$(r_0^2ab \to \exists x\,[r_0^2xb]) \to (\forall y\,[r_0^2ay] \to \exists x\,[r_0^2xb])$	m.p.
(8)	$\forall y\,[r_0^2ay] \to \exists x\,[r_0^2xb]$	m.p.
(9)	$\exists x\,[\forall y\,[r_0^2xy]] \to \exists x\,[r_0^2xb]$	3)
(10)	$\exists x\,[\forall y\,[r_0^2xy]] \to \forall y\,[\exists x[r_0^2xy]]$	4)

Substitutions in provable sentences Given a provable sentence A and a simultaneous substitution s (*a fortiori*, a simple substitution (a/b)), is $s(A)$ again a provable sentence? The answer is in the affirmative; to be precise, we shall prove by induction on n:

7 PROPOSITION *For any sentence A which has a proof of length n, and any simultaneous substitution s, s(A) is a provable sentence.*

Proof If $n = 1$, A must be an instance of an axiom schema.

If it is an instance of one of the first nine axiom schemata, $s(A)$ is an instance of the same axiom schema.

If $A = (a/x) f^x \to \exists x\,[f^x]$, then

$$s(A) = s\,(a/x) f^x \to s\exists x\,[f^x] = (s(a)/x)\, s(f^x) \to \exists x\,[s(f^x)],$$

and this is again an instance of the same axiom schema.

If A is obtained by modus ponens from B and $B \to A$, and the assertion is assumed true for the latter sentences, then $s(B)$ and $s(B \to A) = s(B) \to s(A)$ are provable sentences, therefore so is $s(A)$.

If $A = \exists x [f^x] \to B$ is obtained from $A' = (a/x) f^x \to B$ where $a \notin I_{B,f^x}$ and the assertion is assumed true for A', let $I_{A'} = \{a_1, \dots, a_n\}$ and choose an individual b different from $s(a_1), \dots, s(a_n)$ which has no occurrence in A'. Define s' by

$$s'(c) = s(c) \quad \text{if} \quad c \neq b, \quad s'(b) = b.$$

Then $$s(A) = s'(A) = \exists x [s'(f^x)] \to s'(B);$$

$$(b/a) A' = (b/a)(a/x) f^x \to B = (b/x) f^x \to B,$$

$$s'(b/a) A' = s'(b/x) f^x \to s'(B) = (b/x) s'(f^x) \to s'(B),$$

and the latter sentence is provable by the induction hypothesis, since $s'(B) \rangle b \langle$ and $s'(f^x) \rangle b \langle$. Thus $s(A)$ is provable.

Note that the converse of this proposition is not valid; however, for the moment we cannot give a counter-example, since we do not yet have a criterion for nonprovability of a sentence (this will be done in the next chapter).

Remark In general, if $A_1, \dots, A_n$ is a formal proof of $A_n = A$, then $s(A_1), \dots, s(A_n)$ need not be a proof of $s(A)$; however, this is the case if s is one-to-one, as we now proceed to prove.

If A_i is an instance of an axiom schema, so is $s(A_i)$.

If A_i is preceded by A_j and $A_h = A_j \to A_i$, then $s(A_i)$ is preceded by $s(A_j)$ and $s(A_h) = s(A_j) \to s(A_i)$.

If $A_i = \exists x [f^x] \to B$ is preceded by $A_j = (a/x) f^x \to B$ and $a \notin I_{B,f^x}$, then $s(A_i) = \exists x [s(f^x)] \to s(B)$ is preceded by

$$\begin{aligned} s(A_j) &= s(a/x) f^x \to s(B) \\ &= (s(a)/x)\, s(f^x) \to s(B), \end{aligned}$$

and $s(a)$ occurs neither in $s(B)$ nor in $s(f^x)$, since s is one-to-one.

We have in fact proved a slightly more general result: It is sufficient to assume s one-to-one on the set $\bigcup_1^n I_{A_i}$; i.e., if a has at least one occurrence in at least one sentence A_i, then for any individual b we have $s(a) = s(b) \Rightarrow a = b$. As an application, we prove:

8 Proposition *If A is a provable sentence and I^* is any infinite subset of I such that $I^* \supset I_A$, there is a formal proof of A whose sentences contain no individuals other than those of I^*.*

Proof Let $A_1, \ldots, A_n$ be any proof.

Let $I_A = \{a_1, \ldots, a_p\}$, and let $\{b_1, \ldots, b_q\}$ be the set of individuals with at least one occurrence in a sentence A_i but no occurrences in A. Choose $q + 1$ different individuals $\{c_1, \ldots, c_q, c_{q+1}\}$ in I^*, different from the b_i and the a_j.

Define $s: I \hookrightarrow I^*$ as follows:

$$s(a_j) = a_j,$$

$$s(b_i) = c_i,$$

$$s(d) = c_{q+1} \quad \text{for any other individual } d.$$

Then, by the preceding remark, the sequence $s(A_1), \ldots, s(A_n)$ is a formal proof of $s(A) = A$.

Another interesting property is the following:

9 Proposition *Let A be a provable sentence; then there is a proof of A whose sentences involve no predicates other than those occurring in A.*

Proof Let $A_1, \ldots, A_n = A$ be a proof of A, and assume there is a predicate r^p with occurs in at least one of the sentences A_i, $i \leqslant n - 1$, but has no occurrence in A. Let r_0^q be a fixed predicate which has an occurrence in A, and let u_0 be the atomic sentence $u_0 = r_0^q\, a_1 \cdots a_q$, where the individuals a_i have no occurrence in any sentence A_i $(i \leqslant n)$.

For any sentence B, let B^* denote the expression obtained from B by replacing all sub-expressions of the form $r^p \sigma_1 \cdots \sigma_p$ (where the σ_j are individuals or variables) by u_0.

By induction on the length of a formation sequence for B, it is easy to see that B^* is also a sentence:

If $B \in \mathfrak{a}$ then $B^* = B$ or $B^* = u_0$.

If $B = \neg C$ then $B^* = \neg C^*$, and C^* is a sentence by assumption.

If $B = CkD$ then $B^* = C^* k D^*$.

If $B = Qx\,[(x/a)\,A]$, choose an individual b different from $a_1, \ldots, a_q$ and having no occurrence in A; then

$$B = Qx\,[(x/b)\,(b/a)\,A] = Qx\,[(x/b)\,A'], \quad \text{where} \quad A' = (b/a)\,A;$$

thus $B^* = Qx\,[(x/b)\,A'^*]$.

We claim that the sequence $A_1^*, \ldots, A_{n-1}^*, A_n^*$ is a formal proof of $A = A_n^*$. Indeed, consider A_i^*.

If A_i is an instance of an axiom schema, then:

For the first nine schemata, A_i^* is clearly an instance of the same schema.

If $A_i = (a/x) f^x \rightarrow \exists x [f^x]$, where $f^x = (x/b) B$, choose an individual c different from $a_1, \ldots, a_q$, which has no occurrence in B. Then

$$f^x = (x/c)(c/b) B = (x/c) B', \quad \text{where} \quad B' = (c/b) B,$$

and $(a/x) f^x = (a/x)(x/c) B' = (a/c) B'$. Thus

$$A_i^* = (a/c) B'^* \rightarrow \exists x [(x/c) B'^*].$$

If A_i is obtained by modus ponens from A_j and $A_j \rightarrow A_i$, then A_i^* is obtained by modus ponens from A_j^* and $A_j^* \rightarrow A_i^*$.

If $A_i = \exists x [f^x] \rightarrow B$ and there exists $j < i$ such that $A_j = (a/x) f^x \rightarrow B$ where $a \notin I_{B,f^x}$, let $f^x = (x/c)\, C$ where c is different from $a_1, \ldots, a_q$ (as before). Then $A_i^* = \exists x [(x/c)\, C^*] \rightarrow B^*$ and $A_j = (a/x)(x/c)\, C \rightarrow B = (a/c)\, C \rightarrow B$. Thus $A_j^* = (a/c)\, C^* \rightarrow B^* = (a/x)(x/c)\, C^* \rightarrow B^*$; $(x/c)\, C^* \rangle a \langle$ and $B^* \rangle a \langle$, since a (which has no occurrence in A_j) is different from $a_1, \ldots, a_q$.

To complete the proof, apply the preceding argument to every predicate not occurring in A.

Sublogics We shall now introduce a notion which is of great importance in the sequel, especially in the study of completeness.

Let I^* be an infinite subset of I. A new predicate calculus may be constructed in which I^* is the set of individuals, retaining the same variables, predicates, and connectives as before. We shall call the resulting logic a sublogic. Let E^* denote the set of sentences and T^* the set of provable sentences of this sublogic. Employing a slight abuse of language, we shall call E^* a sublogic of E based on I^*. Conversely, one can also adjoin to I a set I' of new symbols and thus define a "superlogic" E^+ of E based on $I^+ = I \cup I'$.

We state the following results concerning sublogics E^*.

10 Proposition *Let $E(I^*)$ denote the set of sentences $A \in E$ such that $I_A \subset I^*$. Then $E^* = E(I^*)$ and $T^* = T \cap E(I^*)$.*

Proof For the first equality:

Let $A \in E^*$, and reason by induction on the order of A.

If A is an atomic sentence $r_n^p a_1 \cdots a_p$, where $a_1, \ldots, a_p$ are in I^*, then $A \in E(I^*)$.

If $A = \neg B$ or $A = BkC$, where B and C are in $E(I^*)$ by the induction hypothesis, note that $I_A = I_B$ or $I_A = I_B \cup I_C$, so that $I_A \subset I^*$ and $A \in E(I^*)$.

If $A = Qx\,[(x/a)\,B]$ where $a \in I^*$ and $B \in E(I^*)$, it is clear that $A \in E(I^*)$.

Conversely, let $A \in E(I^*)$.

If A is an atomic sentence, clearly $A \in E^*$.

If $A = \neg B$ or $A = BkC$ the result is again trivial.

If $A = Qx\,[(x/a)\,B]$, then:

if $a \in I^*$ then $B \in E(I^*)$, and by the induction hypothesis $B \in E^*$, so that $A \in E^*$.

If $a \notin I^*$, choose $a' \in I^*$ such that $a' \notin I_B$, then:

$(x/a)\,B = (x/a')\,B'$ where $B' = (a'/a)\,B \in E(I^*)$, whence $A \in E^*$.

For the second equality:

If $A \in T^*$, we reason by induction on the length of a proof of A in E^*.

If A is an instance of an axiom schema in E^*, it is also an instance of an axiom schema in E.

If there exists $B \in E^*$ such that $B \in T^*$ and $B \rightarrow A \in T^*$, then by assumption $B \in T$ and $B \rightarrow A \in T$, so that $A \in T$.

If $A = \exists x\,[f^x] \rightarrow B$ where $(a/x) f^x \rightarrow B \in T^*$ or $a \in I^*$ and $a \notin I_{B,f^x}$, then by assumption $(a/x) f^x \rightarrow B \in T$, so that $A \in T$.

If $A = B \rightarrow \forall x\,[f^x]$, the proof is similar.

Conversely, let $A \in T \cap E(I^*)$. Since we have shown that there is a proof whose sentences contain only individuals of I^*, a simple induction proof shows that $A \in T^*$.

Conclusion We have defined a logical calculus which is far more powerful than the propositional calculus, since it permits analysis of propositions on the basis of the individuals they involve. Indeed, the propositional calculus may be regarded as a special case of the predicate calculus in which the only predicates are of weight zero. Moreover, the axioms and rules of inference of the predicate calculus include those of the propositional calculus as special cases. Thus all sentences of the predicate calculus which have the same form as provable sentences of the propositional calculus are also provable in the predicate calculus. Example: $A \vee \neg A$, where A is any sentence of the predicate calculus.

We have simplified presentation of the predicate calculus by directly introducing the concept of axiom schemata, which makes it possible to omit the rule of substitution for sentences. However, we could equally well have adopted a more detailed exposition, in complete analogy with the propositional calculus.

Whenever possible we have adopted the same terminology and notation for the calculus L'' as for the calculus L'. In the sequel this will enable us to state and prove several results which are valid for both calculi.

CHAPTER II

Semantic approach: problems of consistency

1 The concept of interpretation

HITHERTO WE HAVE discussed a purely formal calculus. We have dwelt neither on the various meanings attributable to the objects considered (sentences, provable sentences, etc.) nor on the possible applications thereof. In fact, many different interpretations are possible, e.g., electrical circuits as an interpretation of L'; however, we obviously have in mind a "natural" interpretation, in which sentences represent certain assertions expressed in natural language, and provable sentences assertions that are always true (in the intuitive sense). For example:

The assertion "Whenever it rains I don't go out" may be represented by a sentence of the form $A \to \neg B$.

The assertion "If whenever it rains I don't go out, then when I go out it's not raining" may be represented by a provable sentence of the form $(A \to \neg B) \to (B \to \neg A)$.

In an interpretation of this type the elements of $\mathfrak{a}$, i.e., the atomic sentences of L' and L'', thus play the role of independent variables; in other words, they may be assumed true or false at will. Every sentence may thus be regarded as a function of these variables and represents an assertion whose truth value (true or false) depends on the truth values assigned to the variables. When this truth value is the constant "true" it will follow that the sentence is provable. At any rate, this is one of the conditions imposed on an interpretation. We shall see that this leads to the very important problems of consistency and (later) completeness.

The simplest interpretation is thus to assign to each atomic sentence u a "truth value" λ from a set of two elements $\{0, 1\}$ (where 0 represents "false" and 1 "true"), and then to extend this assignment to the set of all sentences by means of natural operations on 0 and 1, corresponding to the intuitive meaning of the connectives. For example:

$\neg A$ is true if and only if A is false

$A \wedge B$ is true if and only if A and B are true

$A \vee B$ is true if and only if at least one of A and B is true (note that the word "or" is not understood in its exclusive sense); and so on.

The interpretation of quantifiers also easily suggests itself:

$\forall x\, [f^x]$ is true if and only if $(a/x) f^x$ is true for all individuals a; the truth value of this sentence may thus be conceived as the "lowest upper bound" of the truth values of all sentences $(a/x) f^x$.

We shall find it more convenient to consider the concept of interpretation in a broader sense:

An *interpretation* is an ordered pair $(i, \mathbb{V})$, where $\mathbb{V}$ is a nonempty set and i a mapping of $\mathfrak{a}$ into $\mathbb{V}$ satisfying the following conditions:

(1) No conditions are imposed on i.

(2) The set $\mathbb{V}$ is closed under a unary operation, denoted $\neg$, and four binary operations, denoted $\wedge$ $\vee$ $\rightarrow$ $\leftrightarrow$; these operations are defined everywhere, and are otherwise arbitrary.

For the calculus L', no other condition is needed.

For the calculus L'', we impose an additional condition:

(3) $\mathbb{V}$ is a complete lattice with respect to a certain order relation.

Remark It would be sufficient to consider a finite set $\mathbb{V}$; this avoids difficulties connected with set theory and the order relation.

In practice, the mapping i itself is called the interpretation, while $\mathbb{V}$ is called the *value set*. In this terminology, we have the following result:

1 PROPOSITION *Let i be a given interpretation (with a given value set). Then i is uniquely extendible to a mapping $\bar{i}$ of E into $\mathbb{V}$ which satisfies the following conditions for any sentences:*

$$
\begin{aligned}
&(1) \quad \bar{i}\,(\neg A) = \neg \bar{i}(A);\\
&(2) \quad \bar{i}\,(A \wedge B) = \bar{i}(A) \wedge \bar{i}(B);\\
&(3) \quad \bar{i}\,(A \vee B) = \bar{i}(A) \vee \bar{i}(B);\\
&(4) \quad \bar{i}\,(A \rightarrow B) = \bar{i}(A) \rightarrow \bar{i}(B);\\
&(5) \quad \bar{i}\,(A \leftrightarrow B) = \bar{i}(A) \leftrightarrow \bar{i}(B);
\end{aligned}
$$

and for the calculus L'' also

$$
\begin{aligned}
&(6) \quad \bar{i}\,(\exists x\, [f^x]) = \sup_a \bar{i}\,((a/x) f^x);\\
&(7) \quad \bar{i}\,(\forall x\, [f^x]) = \inf_a \bar{i}\,((a/x) f^x).
\end{aligned}
$$

Proof Let E_n denote the set of sentences of order at most n; we prove the following assertion by induction on n:

There exists a unique mapping i_n of E_n into $\mathbb{V}$ which extends i and satisfies the above five [seven] conditions on E_n.

The assertion is true for $n = 0$: take $i_0 = i$, since $E_0 = \mathfrak{a}$.

Assume the assertion true for n; we define a mapping i_{n+1} of E_{n+1} into $\mathbb{V}$ as follows:

$$\begin{aligned}
i_{n+1}(A) &= i_n(A) && \text{if } A \in E_n \\
&= \neg i_n(B) && \text{if } A = \neg B \\
&= i_n(B) \wedge i_n(C) && \text{if } A = B \wedge C \\
&= i_n(B) \vee i_n(C) && \text{if } A = B \vee C \\
&= i_n(B) \rightarrow i_n(C) && \text{if } A = B \rightarrow C \\
&= i_n(B) \leftrightarrow i_n(C) && \text{if } A = B \leftrightarrow C \\
&= \sup_a i_n((a/x)f^x) && \text{if } A = \exists x\,[f^x] \\
&= \inf_a i_n((a/x)f^x) && \text{if } A = \forall x\,[f^x]
\end{aligned}$$

(the last seven lines: when A is of order $n+1$; the last two: only for L'')

i_{n+1} is well defined on E_{n+1}, extends i_n and therefore also i, and satisfies the required conditions on E_{n+1}; for example:

$$\begin{aligned}
i_{n+1}(\neg A) &= i_n(\neg A) && \text{if } A \text{ is of order } \leqslant n-1 \\
&= \neg i_n(A) && \text{by the induction hypothesis;} \\
&= \neg i_n(A) && \text{by definition if } A \text{ is of order } n.
\end{aligned}$$

Since in all cases $i_n(A) = i_{n+1}(A)$, it follows that $i_{n+1}(\neg A) = \neg i_{n+1}(A)$. Another example:

$$\begin{aligned}
i_{n+1}(\exists x\,[f^x]) &= i_n(\exists x\,[f^x]) && \text{if } \exists x\,[f^x] \text{ is of order } \leqslant n \\
&= \sup_a i_n((a/x)f^x) && \text{by the induction hypothesis;} \\
&= \sup_a i_n((a/x)f^x) && \text{by definition if } \exists x\,[f^x] \text{ is of order } n+1;
\end{aligned}$$

and in both cases $i_n((a/x)f^x) = i_{n+1}((a/x)f^x)$.

All other cases are treated similarly.

Finally, i_{n+1} is unique since it must extend i_n and satisfy the above conditions.

Thus the assertion is true for all n.

We now define $\bar{\imath}$ by $\bar{\imath}(A) = i_n(A)$, where n is the order of A. $\bar{\imath}$ is well defined on E; it extends i, for if $u \in \mathfrak{a}$ then $\bar{\imath}(u) = i_0(u) = i(u)$. It also satisfies the required conditions; for example:

$$\begin{aligned}
\bar{\imath}(A \wedge B) &= i_n(A \wedge B) && \text{where } n \text{ is the order of } A \wedge B \\
&= i_n(A) \wedge i_n(B) \\
&= i_p(A) \wedge i_q(B) && \text{where } A \text{ is of order } p \text{ and } B \text{ of order } q \\
&= \bar{\imath}(A) \wedge \bar{\imath}(B).
\end{aligned}$$

Another example:

$$\begin{aligned}\bar{\imath}\,(\forall x\,[f^x]) &= i_n\,(\forall x\,[f^x]) && \text{if } n \text{ is the order of } \forall x\,[f^x]\\ &= \inf_a i_n\,((a/x)\,f^x)\\ &= \inf_a i_{n-1}\,((a/x)\,f^x) && \text{since } (a/x)\,f^x \text{ is of order } n-1\\ &= \inf_a \bar{\imath}\,((a/x)\,f^x).\end{aligned}$$

Finally, $\bar{\imath}$ is unique, since if i' is any other mapping with the same properties the restriction of i' to E_n must coincide with i_n (uniqueness of i_n), so that for every statement A of order n we have $\bar{\imath}(A) = i_n(A) = i'(A)$.

COROLLARY *Given a value set* $\mathbb{V}$, *any mapping* $\bar{\imath}$ *of* E *into* $\mathbb{V}$ *which satisfies the five (seven) conditions of Proposition 1 is uniquely determined by its values on* $\mathfrak{a}$.

We now return to the simplest interpretation, described at the beginning of this chapter.

Systems of truth values

As the value set we take a set $\mathbb{U}$ of two elements 0 and 1, and define the required operations in this set by the following tables (λ and μ represent arbitrary elements of $\mathbb{U}$):

λ	$\neg\lambda$
0	1
1	0

λ	μ	$\lambda \wedge \mu$	$\lambda \vee \mu$	$\lambda \rightarrow \mu$	$\lambda \leftrightarrow \mu$
0	0	0	0	1	1
0	1	0	1	1	0
1	0	0	1	0	0
1	1	1	1	1	1

We define the order relation by $0 < 1$.

Note that $\lambda \leqslant \mu$ if and only if $\lambda \rightarrow \mu = 1$,
$\lambda = \mu$ if and only if $\lambda \leftrightarrow \mu = 1$.

Remark The structure thus defined on $\mathbb{U}$ is a special case of a very general structure—that of a Boolean ring, which we shall use later.

A *system of truth values* is any mapping h of $\mathfrak{a}$ into $\mathbb{U}$, in other words, a particular interpretation. We denote by $\vec{\mathfrak{a}}$ the set of systems of truth values.

By Proposition 1, every system of truth values h has a unique extension $\bar{h}$ to E which satisfies the required conditions. Given a sentence A, we shall call $\tilde{h}(A)$ the *h-truth value* of A; A will be called an *h-true* sentence if $\tilde{h}(A) = 1$, an *h-false* sentence if $\tilde{h}(A) = 0$.

A sentence A is said to be *universally valid** if $\tilde{h}(A) = 1$ for any $h \in \vec{\mathfrak{a}}$; we denote this fact by $\models A$.

The following assertions are immediate:

$\neg A$ is h-true if and only if A is h-false.
$A \wedge B$ is h-true if and only if A and B are h-true.
$A \vee B$ is h-true if and only if at least one of A and B is h-true.
$A \rightarrow B$ is h-true if and only if either A is h-false or A and B are h-true.
$A \leftrightarrow B$ is h-true if and only if A and B are either both h-true or both h-false.
$\exists x\, [f^x]$ is h-true if and only if there exists a such that $(a/x) f^x$ is h-true.
$\forall x\, [f^x]$ is h-true if and only if for all a, $(a/x) f^x$ is h-true.

We have thus related the purely formal approach to the intuitive interpretation in terms of assertions in natural language.

The fundamental problem which now arises is to relate the notion of a provable sentence to that of a universally valid sentence.

As far as the calculus L' is concerned, the following proposition is easily proved:

2 Proposition (Calculus L') *Every provable sentence is universally valid.*

Proof By induction on the length of a formal proof of the provable sentence A.

If A is an instance of an axiom schema, it suffices to verify nine formulas in $\mathbb{U}$ for any λ and μ:

$$(1)\quad \lambda \rightarrow (\mu \rightarrow \lambda) = 1, \quad \text{since if } \lambda = 0 \text{ then } 0 \rightarrow (\mu \rightarrow 0) = 1, \quad \text{and if } \lambda = 1 \text{ then } 1 \rightarrow (\mu \rightarrow 1) = 1 \rightarrow 1 = 1.$$

$$(2)\quad (\lambda \rightarrow (\lambda \rightarrow \mu)) \rightarrow (\lambda \rightarrow \mu) = 1, \quad \text{since}$$
$$\text{if } \lambda = 0 \text{ then } (0 \rightarrow (0 \rightarrow \mu)) \rightarrow (0 \rightarrow \mu) = (0 \rightarrow 1) \rightarrow (0 \rightarrow \mu) = 1 \rightarrow 1 = 1,$$
$$\text{if } \lambda = 1 \text{ then } (1 \rightarrow (1 \rightarrow \mu)) \rightarrow (1 \rightarrow \mu) = (1 \rightarrow \mu) \rightarrow (1 \rightarrow \mu) = 1, \quad \text{etc.}$$

The reader will have no difficulty in verifying the formulas corresponding to the remaining axiom schemata of the propositional calculus.

* *Translator's note* Also known as a *tautology*.

If A is obtained by modus ponens from sentences B and $B \to A$, where by the induction hypothesis $\tilde{h}(B) = \tilde{h}(B \to A) = 1$, then

$$\tilde{h}(B \to A) = \tilde{h}(B) \to \tilde{h}(A) = 1;$$

thus $\tilde{h}(B) \leqslant \tilde{h}(A)$ and $\tilde{h}(A) = 1$.

We now wish to establish the same result for the predicate calculus. Here, however, one case requires particular care. Suppose that $A = \exists x\,[f^x] \to B$ is obtained from $(a/x)f^x \to B$. Reasoning by induction, we assume that

$$\tilde{h}((a/x)f^x \to B) = \tilde{h}((a/x)f^x) \to \tilde{h}(B) = 1.$$

From this we can only conclude that $\tilde{h}((a/x)f^x) \to \tilde{h}(B)$ for the *single* individual a, whereas we wish to establish this inequality for *every* individual b, implying that $\sup_b \tilde{h}((b/x)f^x) \leqslant \tilde{h}(B)$, and hence $\tilde{h}(A) = 1$.

Now the fact that

$$(b/x)f^x \to B = (b/a)\,(a/x)f^x \to (b/a)\,B \quad (\text{where } f^x \rangle a\langle \text{ and } B \rangle a\langle)$$
$$= (b/a)\,((a/x)f^x \to B)$$

suggests studying the effect of substitutions on the evaluation of truth values. This we now proceed to do.

3 Proposition (Calculus L'') *Every provable sentence is universally valid.*

This will result from the following lemma:

Lemma *If $A \in T$, then $\tilde{h}(s(A)) = 1$ for any system of truth values h and any simultaneous substitution s.*

Proof By induction on the length of a formal proof of A.

(1) If A is an instance of an axiom schema, we known that $s(A)$ is also an instance of the same schema. If this schema is one of the first nine, then $\tilde{h}(s(A)) = 1$ as in the propositional calculus L'. If this schema is S10, it suffices to show that $\tilde{h}((a/x)f^x) \leqslant \tilde{h}(\exists x\,[f^x])$, which is trivial since $\tilde{h}(\exists x\,[f^x]) = \sup_b \tilde{h}((b/x)f^x)$. The treatment of S11 is similar.

(2) If A is derived by modus ponens from B and $B \to A$ and the assertion is assumed true for the latter two sentences: $\tilde{h}(s(B)) = 1$ and $\tilde{h}(s(B \to A)) = 1$, i.e., $\tilde{h}(s(B)) \to \tilde{h}(s(A)) = 1$, then $\tilde{h}(s(A)) = 1$.

(3) Now let $A = \exists x\,[f^x] \to B$ be derived from $(a/x)f^x \to B$, where $f^x \rangle a\langle$ and $B \rangle a\langle$, and assume the assertion true for the latter sentence. Let b be any individual and s' the substitution defined by

$$s'(c) = s(c) \quad \text{if} \quad c \neq a, \quad s'(a) = b.$$

Then $s'(B) = s(B)$ since $B \rangle a \langle$. Now $(b/x)\, s(f^x) = s'(a/x) f^x$, so that

$$\tilde{h}\,((b/x)\, s(f^x)) = \tilde{h}\,(s'(a/x) f^x).$$

But, by the induction hypothesis, $\tilde{h}\,(s'((a/x) f^x \to B)) = 1$; in other words,

$$\tilde{h}\,(s'(a/x) f^x) \leqslant \tilde{h}\,(s'(B)) = \tilde{h}\,(s(B)).$$

Hence $\tilde{h}\,((b/x)\, s(f^x)) \leqslant \tilde{h}\,(s(B))$ for any b, and therefore

$$\sup_b \tilde{h}\,((b/x)\, s(f^x)) \leqslant \tilde{h}\,(s(B)),$$

or $\tilde{h}\,(\exists x\, [s(f^x)]) \leqslant \tilde{h}\,(s(B))$. Thus $\tilde{h}\,(s(A)) = 1$.

(4) The case $A = B \to \forall x\, [f^x]$ is treated similarly.

On the basis of the results just established, we can now state the consistency theorems (for both calculi L' and L'').

Consistency theorems

FIRST CONSISTENCY THEOREM (Semantic Consistency) *Every provable sentence is universally valid.*

Remark 1 The converse of this theorem is the Semantic Completeness Theorem.

Remark 2 The theorem provides us with a criterion for non-provability of sentences. It suffices to find a system of truth values h such that $\tilde{h}(A) = 0$.

SECOND CONSISTENCY THEOREM (Syntactic Consistency) *There exist non-provable sentences; in other words, $T \neq E$.*

Proof It suffices to consider the system of truth values h which vanishes identically on $\mathfrak{a}$: $\tilde{h}(u) = 0$; thus no atomic sentence is provable. Similarly, $\neg u$ is not a provable sentence (take h identically equal to 1).

THIRD CONSISTENCY THEOREM (Syntactic Consistency) *There is no sentence A such that $A \in T$ and $\neg A \in T$.*

Proof If $\tilde{h}(A) = 1$, then $\tilde{h}(\neg A) = \neg \tilde{h}(A) = 0$.

Remark 1 The last theorem is extremely important, for it shows that the theories we have constructed are noncontradictory. It is in this sense that we should understand the concept of syntactic consistency, while semantic consistency expresses the agreement of theory with interpretation: only those sentences interpreted as "true" can possibly be proved in the theory (the fact that *all* such sentences are provable is the Semantic Completeness Theorem).

Remark 2 The last two theorems are corollaries of the first; nevertheless, it is interesting to show that they are equivalent.

In fact, Theorem 3 implies Theorem 2 trivially.

Conversely, suppose there exists a sentence A such that $A \in T$ and $\neg A \in T$. Let B be any sentence. We construct a formal proof consisting of the following sentences:

Formal proof of A,	
A,	
formal proof of $\neg A$,	
$\neg A$,	
$\neg A \to (\neg B \to \neg A)$,	S1
$\neg B \to \neg A$,	m.p.
$(\neg B \to \neg A) \to (A \to B)$,	S7
$A \to B$,	m.p.
B.	m.p.

Thus $B \in T$, and this implies $T = E$.

Uniform systems of truth values In the case of the predicate calculus one can introduce a special type of systems of truth values, which yields the Syntactic Consistency Theorems very rapidly and simply (though not the Semantic Consistency Theorem in its general form).

A system of truth values h is said to be *uniform* if, for any atomic sentence u, the value $h(u)$ does not depend on the individuals of u but only on its predicate. This is equivalent to specifying a mapping g of the set of predicates into $\mathbb{U}$ and defining h by

$$h\,(r_n^{(p)} a_1 \cdots a_p) = g(r\,).$$

The following assertion holds for any uniform system of truth values:

4 Proposition *For any sentence A and any simultaneous substitution s,*

$$\tilde{h}\,(s(A)) = \tilde{h}(A).$$

Proof By induction on the order of A.

If $A \in \mathfrak{a}$ the assertion is true by definition.

If $A = \neg B$ or $A = BkC$ the proof is trivial.

If $A = \exists x\,[f^x]$ and $f^x = (x/a)\,B$:

$$\tilde{h}(A) = \sup_b \tilde{h}\,((b/x)\,f^x);$$

but

$$\tilde{h}\,((b/x)\,f^x) = \tilde{h}\,((b/x)\,(x/a)\,B) = \tilde{h}\,((b/a)\,B) = \tilde{h}(B),$$

and hence $\tilde{h}(A) = \tilde{h}(B)$.

Now $s(A) = \exists x\,[s(f^x)]$ and $\tilde{h}\,(s(A)) = \sup_b \tilde{h}\,((b/x)\,s(f^x))$. Fix b and choose an individual $c \notin I_{fx}$; define s' by

$$s'(d) = s(d) \quad \text{if} \quad d \neq c,$$

$$s'(c) = b.$$

Then $(b/x)\,s(f^x) = s'((c/x)\,f^x) = s'((c/a)\,B)$, and thus $\tilde{h}\,((b/x)\,s(f^x)) = \tilde{h}\,(s'(c/a)\,B) = \tilde{h}(B)$. Hence $\tilde{h}\,(s(A)) = \tilde{h}(B) = \tilde{h}(A)$.

If $A = \forall x\,[f^x]$ the proof is similar.

Note that $\tilde{h}\,(\exists x\,[f^x]) = \tilde{h}\,(\exists x\,[f^x]) = h\,((a/x)\,f^x)$ for any individual a. We can now prove:

5 Proposition *If $A \in T$ then $\tilde{h}(A) = 1$ for any uniform system of truth values.*

Proof If A is an instance of an axiom schema, or if A is obtained by modus ponens, the proof is easy.

If $A = \exists x\,[f^x] \to B$ is obtained from $(a/x)\,f^x \to B$, where $a \notin I_{B,f^x}$, then the induction hypothesis,

$$\tilde{h}\,((a/x)\,f^x \to B) = \tilde{h}\,((a/x)\,f^x) \to \tilde{h}(B) = 1,$$

implies that $\tilde{h}\,(\exists x\,[f^x]) \to \tilde{h}(B) = \tilde{h}(A) = 1$.

Syntactic consistency now results from the fact that for every atomic sentence u one can find a uniform system of truth values h such that $h(u) = 0$.

Problems of independence

Another problem related to the concept of interpretation is that of the independence of the axioms, or rather the axiom schemata. This problem is not of major importance, and there is no inconvenience in working with superfluous axioms (this even facilitates certain arguments). It is only for reasons of "elegance" that one is at all interested in independent axioms.

To simplify matters we shall only consider the problem for the propositional calculus, leaving most of the proofs to the reader.

We first introduce a few definitions.

Consider an axiom schema $S_m = \langle A, B, C\rangle_m$, and let $\mathscr{A}_{-m}$ be the set of all the remaining axiom schemata.

A formal Δ_{-m}-proof is any sequence of sentences, each of which is either an instance of an axiom schema of $\mathscr{A}_{-m}$ or derived from two of its predecessors by modus ponens.

We shall call S_m an *independent axiom schema* if there is an instance of S_m which has no formal Δ_{-m}-proof.

A set of axiom schemata is said to be independent if each axiom schema is independent.

Remark It may happen that an axiom schema S_m is independent, but certain instances of S_m possess a formal Δ_{-m}-proof.

Example $(A \leftrightarrow A) \to (A \to A)$ is an instance of S4, but it is also an instance of S5.

Example $(A \leftrightarrow B) \to (A \to B)$ is an instance of S4;
$(A \to B) \to ((B \to A) \to (A \leftrightarrow B))$ is an instance of S6;
$((A \leftrightarrow B) \to (A \to B)) \to (((A \to B) \to ((B \to A) \to (A \leftrightarrow B))) \to ((A \leftrightarrow B) \to ((B \to A) \to (A \leftrightarrow B))))$ is an instance of S3.

Now by two applications of modus ponens we get

$$(A \leftrightarrow B) \to ((B \to A) \to (A \leftrightarrow B)),$$

which is also an instance of S1.

To show that an axiom schema S_m is independent, we proceed in the following manner.

We wish to find an interpretation $(i, \mathbb{V})$ satisfying the following conditions:

(1) There exists a proper nonempty subset $\mathbb{D}$ of $\mathbb{V}$ which is closed with respect to modus ponens, i.e., if λ and $\lambda \to \mu$ belong to $\mathbb{D}$ then μ belongs to $\mathbb{D}$; we call $\mathbb{D}$ the set of *designated elements* of $\mathbb{V}$.

(2) For any instance of an axiom schema of $\mathscr{A}_{-m}$, $\bar{\imath}(\langle A, B, C\rangle)$ is a designated element.

(3) There exists at least one instance of S_m such that $\bar{\imath}(\langle A, B, C\rangle_m)$ is not a designated element.

If there exists such an interpretation, it follows that S_m is independent, since it is easily seen by induction that, for any sentence having a formal Δ_{-m}-proof, $\bar{\imath}(A) \in \mathbb{D}$.

It would take too long to verify the independence of the nine axiom schemata of the propositional calculus (though this can be done). We confine ourselves to the following case.

Independence of S1 The value set will consist of three elements:

$$\mathbb{V} = \{0, 1, 2\}.$$

The operation $\rightarrow$ is defined as follows:

$$\begin{array}{lll} 0 \rightarrow 0 = 1 & 1 \rightarrow 0 = 0 & 2 \rightarrow 0 = 0 \\ 0 \rightarrow 1 = 1 & 1 \rightarrow 1 = 1 & 2 \rightarrow 1 = 0 \\ 0 \rightarrow 2 = 1 & 1 \rightarrow 2 = 1 & 2 \rightarrow 2 = 1 \end{array}$$

and the other operations will be defined below.

The set of designated elements will be $\mathbb{D} = \{1, 2\}$; it is easily seen to be closed with respect to modus ponens.

Now consider $u \rightarrow (v \rightarrow u)$, where u and v are different atomic sentences, which is an instance of S1. Choose i such that $i(u) = 2$ and $i(v) = 0$. Then $2 \rightarrow (0 \rightarrow 2) = 2 \rightarrow 1 = 0$, which is not a designated element.

On the other hand, every instance of the other axiom schemata assumes a designated value:

S2) The value has the form $(\lambda \rightarrow (\lambda \rightarrow \mu)) \rightarrow (\lambda \rightarrow \mu)$;

$$\begin{array}{ll} \text{if } \lambda = 0: & (0 \rightarrow (0 \rightarrow \mu)) \rightarrow (0 \rightarrow \mu) = 1 \rightarrow 1 = 1; \\ \text{if } \lambda = 1 \text{ and } \mu = 0: & (1 \rightarrow (1 \rightarrow 0)) \rightarrow (1 \rightarrow 0) \\ & = (1 \rightarrow 0) \rightarrow (1 \rightarrow 0) = 1; \\ \text{if } \lambda = 1 \text{ and } \mu = 1: & (1 \rightarrow (1 \rightarrow 1)) \rightarrow (1 \rightarrow 1) = 1; \\ \text{if } \lambda = 1 \text{ and } \mu = 2: & (1 \rightarrow (1 \rightarrow 2)) \rightarrow (1 \rightarrow 2) = 1; \\ \text{if } \lambda = 2 \text{ and } \mu = 0 \text{ or } 1: & (2 \rightarrow 0) \rightarrow 0 = 0 \rightarrow 0 = 1; \\ \text{if } \lambda = 2 \text{ and } \mu = 2: & (2 \rightarrow 1) \rightarrow 1 = 0 \rightarrow 1 = 1. \end{array}$$

S3) $(\lambda \rightarrow \mu) \rightarrow ((\mu \rightarrow \nu) \rightarrow (\lambda \rightarrow \nu))$:

If $\lambda = 0$ then $\lambda \rightarrow \mu = \lambda \rightarrow \nu = 1$, thus we have $1 \rightarrow ((\mu \rightarrow \nu) \rightarrow 1)$ and since $\mu \rightarrow \nu = 0$ or 1, the result is 1 in all cases;

$$\begin{array}{rl} \text{if } \lambda = 1 \text{ and } \mu = 0: & 0 \rightarrow ((\quad)) = 1; \\ \lambda = 1 \text{ and } \mu = 1: & 1 \rightarrow ((1 \rightarrow \nu) \rightarrow (1 \rightarrow \nu)) = 1 \rightarrow 1 = 1; \\ \lambda = 1 \text{ and } \mu = 2: & 1 \rightarrow ((2 \rightarrow \nu) \rightarrow (1 \rightarrow \nu)) = 1 \text{ in all cases}; \\ \text{if } \lambda = 2 \text{ and } \mu = 0: & 0 \rightarrow ((\quad)) = 1; \\ \lambda = 2 \text{ and } \mu = 1: & 0 \rightarrow ((\quad)) = 1; \\ \lambda = 2 \text{ and } \mu = 2: & 1 \rightarrow ((2 \rightarrow \nu) \rightarrow (2 \rightarrow \nu)) = 1 \rightarrow 1 = 1. \end{array}$$

S4) and S5) We define the operation $\leftrightarrow$ by

$$\lambda \leftrightarrow \mu = 1 \quad \text{if} \quad \lambda = \mu, \quad \lambda \leftrightarrow \mu = 0 \quad \text{if} \quad \lambda \neq \mu.$$

The values of S4) and S5) are then:

$$\begin{aligned} &\text{If } \lambda = \mu\text{:} \quad 1 \to 1 = 1; \\ &\text{if } \lambda \neq \mu\text{:} \quad 0 \to (\quad) = 1. \end{aligned}$$

S6) $(\lambda \to \mu) \to ((\mu \to \lambda)) \to (\lambda \leftrightarrow \mu)$:

If $\lambda = \mu$: $1 \to (1 \to 1) = 1$;

if $\lambda \neq \mu$: $(\lambda \to \mu) \to ((\mu \to \lambda) \to 0)$ has the following value:

$\lambda = 0$ and $\mu = 1$: $1 \to (0 \to 0) = 1$;

$\mu = 2$: $1 \to (0 \to 0) = 1$;

$\lambda = 1$ and $\mu = 0$: $0 \to (\quad) = 1$;

$\mu = 2$: $1 \to (0 \to 0) = 1$;

$\lambda = 2$ and $\mu = 0$ or 1: $0 \to (\quad) = 1$.

S7) We define the operation $\neg$ by: $\neg 0 = 2$, $\neg 2 = 0$, $\neg 1 = 1$. Then $\lambda \to \mu = \neg\mu \to \neg\lambda$ in all cases; in fact:

$$\begin{aligned} &\text{if } \lambda = 0\text{:} && 0 \to \mu = 1 \quad \text{and} \quad \neg\mu \to 2 = 1; \\ &\text{if } \lambda = 1\text{:} && \mu = 0\text{:} \quad 1 \to 0 = 0 = 2 \to 1; \\ &\quad \mu = 1\text{:} && 1 \to 1 = 1 = 1 \to 1; \\ &\quad \mu = 2\text{:} && 1 \to 2 = 1 = 0 \to 1; \\ &\text{if } \lambda = 2\text{:} && \mu = 0\text{:} \quad 2 \to 0 = 0 = 2 \to 0; \\ & && \mu = 1\text{:} \quad 2 \to 1 = 0 = 1 \to 0; \\ & && \mu = 2\text{:} \quad 2 \to 2 = 1 = 0 \to 0. \end{aligned}$$

S8) and S9) We define the operations $\vee$ and $\wedge$ by:

$$\lambda \vee \mu = \neg\lambda \to \mu, \quad \lambda \wedge \mu = \neg(\neg\lambda \vee \neg\mu).$$

We have thus proved that the axiom schema S1 is independent in the calculus L'. This procedure is very lengthy since it requires a considerable amount of checking, as is evident from the one example we have worked out. However, it involves no essential difficulties, apart perhaps from the need for a little imagination in finding a suitable interpretation for each axiom schema.

CHAPTER III

Set-theoretic approach: concepts of deducibility

THIS CHAPTER DEALS with both calculi L' and L''.

1 Different concepts of deducibility

In the preceding chapters we defined the concept of a provable sentence, or, so to speak, the concept of "absolute truth" (with respect to a primitive axiom system). We are now about to extend this concept and define "relative truth", that is to say, the concept of a sentence which is true under certain additional assumptions; this will provide us with the means for further study of certain sets of sentences which play a fundamental role.

First concept

Sentence deducible from another sentence A sentence B is said to be *deducible* from another sentence A if $A \to B \in T$. Notation: $A \vdash B$.

We denote the set of sentences deducible from A by $T(A)$. In paiticular:

1 PROPOSITION *If $A \in T$ then $T(A) = T$.*

Proof If $A \in T$ and $A \to B \in T$, then $B \in T$ (by modus ponens).

Conversely, if $B \in T$ then $A \vdash B$ for any sentence A; this follows from the following formal proof:

$$\ldots, B \quad \text{(formal proof of } B)$$

$$B \to (A \to B)$$

$$A \to B.$$

Note that the relation $A \vdash B$ defines a *partial order* in E: it is reflexive, $A \vdash A$, since $A \to A \in T$; and it is transitive—for if $A \vdash B$ and $B \vdash C$, consider the following formal proof:

$$\ldots, A \to B \quad \text{(formal proof of } A \to B)$$
$$\ldots, B \to C \quad \text{(formal proof of } B \to C)$$
$$(A \to B) \to ((B \to C) \to (A \to C))$$
$$(B \to C) \to (A \to C), \quad \text{whence} \quad A \to C \in T, \quad \text{so that} \quad A \vdash C.$$

Thus the relation "$A \vdash B$ and $B \vdash A$" is an *equivalence* relation in E.

An equivalent formulation of this relation is $A \leftrightarrow B \in T$.

In fact, if $A \to B \in T$ and $B \to A \in T$, we obtain $A \leftrightarrow B \in T$ from the statement $(A \to B) \to ((B \to A) \leftrightarrow (A \to B))$ by two applications of modus ponens.

If $A \leftrightarrow B \in T$, then since $(A \leftrightarrow B) \to (A \to B) \in T$
and $(A \leftrightarrow B) \to (B \to A) \in T$,

application of modus ponens gives $A \to B \in T$ and $B \to A \in T$.

Let us denote this equivalence relation by ARB. It will play a very important role in the sequel; in fact, we shall see that the quotient set E/R can be given the structure of a Boolean ring.

Note that the set T is an equivalence class, for:

if $A \in T$ and ARB, then $A \vdash B$, so that $B \in T$;
if $A \in T$ and $B \in T$, then $A \vdash B$ and $B \vdash A$, so that ARB.

Second concept

Sentence deducible from hypotheses We shall now generalize the preceding discussion; instead of deducibility from another sentence, we shall define deducibility from a set of sentences.

Let $J \subset E$ be any set of sentences. A *deduction from the hypotheses J* is any finite sequence of sentences $A_1, \ldots, A_n$ each of which satisfies (at least) one of the following conditions:

(1) $A_i \in J$;
(2) $A_i \in T$;
(3) there exist $j < i$ and $h < i$ such that $A_h = A_j \to A_i$.

We call such a sequence a deduction of A_n, of length n, from the hypotheses J.

We now define a sentence A to be *deducible from the hypotheses J* if it has at least one deduction from the hypotheses J; we then write $J \vdash A$, and denote the set of sentences deducible from the hypotheses J by $T(J)$.

Remark 1 $T(J) \supset J \cup T$.

Remark 2 $T(J)$ is closed under modus ponens: if $J \vdash A$ and $J \vdash A \to B$, then $J \vdash B$.

Remark 3 We repeat that this definition is applicable to both the propositional calculus and the predicate calculus (though the rules of inference for the calculus L'' do not figure explicitly in the definition—this question will be made more precise in Remark 2, p. 51f).

This definition of deducibility is indeed a generalization of the first (in which J is a set consisting of a single sentence). To prove this, we first establish the following lemma.

LEMMA *If $A \to C \in T$ and $A \to (C \to B) \in T$, then $A \to B \in T$.*

Proof Consider the following formal proof:

$\ldots, A \to C$ (formal proof of $A \to C$)	
$\ldots, A \to (C \to B)$ (formal proof of $A \to (C \to B)$)	
$(A \to C) \to ((C \to B) \to (A \to B))$	S3
$(C \to B) \to (A \to B)$	m.p.
$(A \to (C \to B)) \to (((C \to B) \to (A \to B)) \to (A \to (A \to B)))$	S3
$((C \to B) \to (A \to B)) \to (A \to (A \to B))$	m.p.
$A \to (A \to B)$	m.p.
$(A \to (A \to B)) \to (A \to B)$	S2
$A \to B.$	m.p.

Now consider the two definitions of deducibility.

If $A \vdash B$ in the sense of the first definition, the sequence $A, A \to B, B$ is a deduction of B from the hypotheses $\{A\}$, so that $\{A\} \vdash B$ in the sense of the second definition.

If $\{A\} \vdash B$ in the sense of the second definition, we prove that $A \vdash B$ by induction on the length of a deduction of B from the hypotheses $\{A\}$:

If $n = 1$, then either $B = A$ or $B \in T$, and in either case $A \to B \in T$.

Assuming the assertion true for deductions of length $\leqslant n - 1$, let B have a deduction of length n. Then:

If $B = A$ or $B \in T$ the assertion is trivial.

If B is obtained by modus ponens from C and $C \to B$, then by the induction hypothesis we have $A \to C \in T$ and $A \to (C \to B) \in T$, and by the lemma $A \to B \in T$.

Thus the two definitions are indeed equivalent. If J consists of a single sentence A, we shall therefore continue to write $A \vdash B$ and $T(A)$ instead of $\{A\} \vdash B$ and $T(\{A\})$.

Of the many properties of the relation $J \vdash A$, we shall study only the most important.

2 PROPOSITION *For any subset* $J \subset T$, $T(J) = T$.

Proof If $A \in T$, then $J \vdash A$ for any J.

If $J \vdash A$, one proves by induction on the length of a deduction of A from the hypotheses J that $A \in T$.

In particular: $T(T) = T$,

$T(\emptyset) = T$, in other words, $\vdash A$ if and only if $\emptyset \vdash A$.

3 PROPOSITION *If* $J \subset J'$, *then* $T(J) \subset T(J')$.

For any deduction from the hypotheses J is *a fortiori* a deduction from the hypotheses J'.

4 PROPOSITION *For any* J, $T(T(J)) = T(J)$.

Proof $J \subset T(J)$, therefore $T(J) \subset T(T(J))$.

If $A \in T(T(J))$, one proves by induction on the length of a deduction of A from the hypotheses $T(J)$ that $A \in T(J)$.

5 PROPOSITION $J \vdash A$ *if and only if there exists a finite subset* J_A^0 *(depending on* A*) of* J *such that* $J_A^0 \vdash A$.

Proof If there exists a finite subset such that $J_A^0 \vdash A$, then $J \vdash A$ (Proposition 3).

On the other hand, if $J \vdash A$, let $A_1, \ldots, A_n = A$ be a deduction of A from the hypotheses J, and put $J_A^0 = \{A_1, \ldots, A_n\} \cap J$. Then $A_1, \ldots, A_n$ is also a deduction from the hypotheses J_A^0.

6 PROPOSITION *If* $J \vdash A \to B$, *then* $J \cup \{A\} \vdash B$; *we represent this by the symbol*

$$\frac{J \vdash A \to B}{J, A \vdash B}$$

In fact, if $A \to B \in T(J)$, then *a fortiori* $A \to B \in T(J, A)$; but $A \in T(J, A)$, and therefore $B \in T(J, A)$.

7 PROPOSITION $\dfrac{J, A \vdash B}{J \vdash A \to B}$. This is the converse of the preceding proposition, and is known as the *Deduction Theorem*.

Proof We reason by induction on the length n of a deduction of B from the hypotheses J, A.

If $n = 1$, there are three possibilities:

If $B \in T$ or $B \in J$, then $J \vdash B$. But $J \vdash B \to (A \to B)$ (axiom S1), so that $J \vdash A \to B$.

If $B = A$, then $J \vdash A \to A$ since $A \to A \in T$.

Assume the assertion true for all deductions of length at most $n-1$. Suppose B is obtained by modus ponens from two preceding sentences C and $C \to B$.

$J, A \vdash C$, so that by the induction hypothesis $J \vdash A \to C$;

$J, A \vdash C \to B$, so that by the induction hypothesis $J \vdash A \to (C \to B)$.

Now the sequence constructed in the lemma on p. 40 is a deduction of $A \to B$ from the hypotheses J.

COROLLARY *$J \vdash A$ if and only if there exists a finite number of sentences $B_1, \ldots, B_n$ in J such that*

$$B_1 \to (B_2 \to (\cdots \to (B_n \to A) \cdots)) \in T.$$

Proof If $J \vdash A$, there exists a finite subset J^0 of J such that $J^0 \vdash A$. Let $J^0 = \{B_1, \ldots, B_n\}$ (the case $J^0 = \emptyset$ is trivial: $A \in T$). Then

$$B_1, \ldots, B_{n-1}, B_n \vdash A,$$

therefore

$$B_1, \ldots, B_{n-1} \vdash B_n \to A, \quad \text{etc.}$$

and finally

$$\emptyset \vdash B_1 \to (B_2 \to (\cdots \to (B_n \to A) \cdots)).$$

Conversely, using Proposition 6 instead of the deduction theorem we get

$$B_1, \ldots, B_n \vdash A,$$

whence *a fortiori* $J \vdash A$.

Remark The above assertion is independent of the order of the sentences B_i. The case $J = \emptyset$ is trivial: $A \in T$.

8 PROPOSITION $\dfrac{J \vdash A,\ A \vdash B}{J \vdash B}$.

For since $A \in T(J)$ we have $\{A\} \subset T(J)$, therefore $T(A) \subset T(J)$, and so $B \in T(J)$.

9 PROPOSITION *For any sentences A, B, C:*

$$A \vdash B \to A$$
$$A \to (A \to B) \vdash A \to B$$
$$A \to B \vdash (B \to C) \to (A \to C)$$
$$A \to B,\ B \to C \vdash A \to C$$
$$A \leftrightarrow B \vdash A \to B$$
$$A \leftrightarrow B \vdash B \to A$$
$$A \to B \vdash (B \to A) \to (A \leftrightarrow B)$$
$$A \to B,\ B \to A \vdash A \leftrightarrow B$$
$$\neg B \to \neg A \vdash A \to B.$$

The proofs of these assertions are immediate, on the basis of axioms S1 to S7 and Proposition 6.

For example, axiom S1 is $\emptyset \vdash A \to (B \to A)$
whence $A \vdash B \to A.$

10 PROPOSITION *For any sentences A and B:*

$$A \vee B \vdash \neg A \to B$$
$$\neg A \to B \vdash A \vee B$$
$$A \wedge B \vdash \neg(\neg A \vee \neg B)$$
$$\neg(\neg A \vee \neg B) \vdash A \wedge B.$$

Here one uses axioms S8 and S9; for example:
S8 may be written $\emptyset \vdash A \vee B \leftrightarrow (\neg A \to B)$;
now $A \vee B \leftrightarrow (\neg A \to B) \vdash A \vee B \to (\neg A \to B)$ (Proposition 9)
so that $\emptyset \vdash A \vee B \to (\neg A \to B)$ (Proposition 8)
and so $A \vee B \vdash \neg A \to B.$

11 PROPOSITION (calculus L'') *For any individual a and any quantifiable formula f^x,*

$$(a/x)f^x \vdash \exists x\,[f^x],$$

$$\forall x\,[f^x] \vdash (a/x)f^x.$$

12 PROPOSITION $\dfrac{J \vdash A \to B,\ J \vdash B \to C}{J \vdash A \to C}.$

Proof $(A \to B) \to ((B \to C) \to (A \to C)) \in T(J)$ (axiom);
thus if $A \to B \in T(J)$, then $(B \to C) \to (A \to C) \in T(J)$,
and if $B \to C \in T(J)$, then $A \to C \in T(J)$.

13 PROPOSITION *For any sentences A, B, C:*

$$A \to (C \to B) \vdash C \to (A \to B).$$

Proof $C \vdash A \to C$ (axiom S1)
$A \to C, A \to (C \to B) \vdash A \to B$ (lemma, p. 40).
Thus $A \to C \vdash (A \to (C \to B)) \to (A \to B)$
whence $C \vdash (A \to (C \to B)) \to (A \to B)$ (Proposition 8),
so that $C, A \to (C \to B) \vdash A \to B$,
or $A \to (C \to B) \vdash C \to (A \to B)$.

14 PROPOSITION $J \vdash A \vee B$ *if and only if* $J, \neg A \vdash B$.
For $A \vee B \vdash \neg A \to B$ and $\neg A \to B \vdash A \vee B$.

15 PROPOSITION *For any sentence B*, $\dfrac{J \vdash \neg A}{J \vdash A \to B}$.

For $\neg A \vdash \neg B \to \neg A$ *and* $\neg B \to \neg A \vdash A \to B$.

16 PROPOSITION $J \vdash A$ *if and only if* $J \vdash \neg\neg A$.

Proof On the one hand:

If $J \vdash \neg\neg A$, then $J \vdash \neg A \to \neg\neg\neg A$ (Proposition 15);

but $\neg A \to \neg\neg\neg A \vdash \neg\neg A \to A$ (S7),

so that $J \vdash \neg\neg A \to A$

and $J \vdash A$ by modus ponens.

In particular,

$$\neg\neg A \vdash \neg\neg A, \quad \text{and therefore} \quad \neg\neg A \vdash A.$$

Now this is true for any sentence A, therefore also $\neg\neg\neg A \vdash \neg A$, or $\emptyset \vdash \neg\neg\neg A \to \neg A$.

But $\neg\neg\neg A \to \neg A \vdash A \to \neg\neg A$, and so $\emptyset \vdash A \to \neg\neg A$, or $A \vdash \neg\neg A$. Thus, if $J \vdash A$ then $J \vdash \neg\neg A$.

17 PROPOSITION $\dfrac{A \vdash B}{\neg B \vdash \neg A}$.

In fact, if $A \vdash B$, then since $\neg\neg A \vdash A$ we have $\neg\neg A \vdash B$; but $B \vdash \neg\neg B$, therefore $\neg\neg A \vdash \neg\neg B$, whence $\neg B \vdash \neg A$ by S7.

Remark In fact, one can prove a more precise result:

$A \to B \vdash \neg B \to \neg A$ ("inversion" of S7). The proof is as follows.

$A \to B \vdash A \to B$

$A \to B, A \vdash B$

$A \to B, A \vdash \neg\neg B$

$A \vdash (A \to B) \to \neg\neg B$

$\neg\neg A \vdash (A \to B) \to \neg\neg B$

$\neg\neg A, A \to B \vdash \neg\neg B$

$A \to B \vdash \neg\neg A \to \neg\neg B$

$\neg\neg A \to \neg\neg B \vdash \neg B \to \neg A$ (by S7)

and hence $A \to B \vdash \neg B \to \neg A$.

18 PROPOSITION $A, B \vdash C$ *if and only if* $A \wedge B \vdash C$.

Proof If $A \wedge B \vdash C$, then $\neg C \vdash \neg(A \wedge B)$;

but $\neg(\neg A \vee \neg B) \vdash A \wedge B$,

therefore $\neg(A \wedge B) \vdash \neg\neg(\neg A \vee \neg B)$, whence $\neg(A \wedge B) \vdash \neg A \vee \neg B$;

$\neg C \vdash \neg A \vee \neg B$

$\neg C, \neg\neg A \vdash \neg B$ (Proposition 14)

$\neg\neg A \vdash \neg C \rightarrow \neg B$

$\neg\neg A \vdash B \rightarrow C$, and hence $A \vdash B \rightarrow C$ or $A, B \vdash C$.

On the other hand, if $A, B \vdash C$, we need only reverse the reasoning.

COROLLARY *Let J be a finite set of sentences $\{A_1, \ldots, A_n\}$, and set*

$$C_1 = A_1$$
$$C_2 = C_1 \wedge A_2$$
$$\cdots\cdots\cdots\cdots$$
$$C_n = C_{n-1} \wedge A_n.$$

Then $J \vdash A$ if and only if $C_n \vdash A$.

Proof By induction on n.

The case $n = 1$ is trivial.

Assume the assertion true for $n - 1$. Then the following assertions are equivalent:

$$A_1, \ldots, A_n \vdash A$$
$$A_1, \ldots, A_{n-1} \vdash A_n \rightarrow A$$
$$C_{n-1} \vdash A_n \rightarrow A$$
$$C_{n-1}, A_n \vdash A$$
$$C_n \vdash A \quad \text{by Proposition 18.}$$

Note that the order in which the sentences A_i appear is immaterial; we can therefore denote the conjunction of the sentences in J, in arbitrary order, by $\wedge J$:

$$\wedge J = ((A_{i1} \wedge A_{i2}) \wedge A_{i3}) \wedge A_{i4} \cdots$$

Then, since J is finite (and nonempty), $J \vdash A$ if and only if $\wedge J \vdash A$.

If J is not finite, then $J \vdash A$ if and only if there exists a finite subset J_A^0 of J such that $\wedge J_A^0 \vdash A$.

In the sequel, we shall use the symbol θ to denote a provable sentence, chosen once and for all, such as $\theta = u^0 \rightarrow u^0$ (where u^0 is a fixed atomic sentence); ω will denote the sentence $\neg\theta = \neg(u^0 \rightarrow u^0)$.

19 PROPOSITION *$J \vdash A$ if and only if $J, \neg A \vdash \omega$ (or $J \vdash A \vee \omega$).*

Proof If $J \vdash A$ then $J \vdash \neg\neg A$

$J \vdash \neg A \rightarrow \omega$

$J, \neg A \vdash \omega$;

and if $J, \neg A \vdash \omega$, then $J \vdash \neg A \to \omega$

$J \vdash \neg A \to \neg\theta$

$J \vdash \theta \to A;$

but $J \vdash \theta$, since $\theta \in T$,

and so $J \vdash A$.

This property gives rise to our third and last generalization of the concept of deducibility:

Third concept

Global deducibility Let J and K be any two sets of sentence. $\neg K$ denotes the set of all sentences $\neg B$, where $B \in K$.

K is said to be *globally deducible from the hypotheses* J, and we write $J \vdash K$, if

$$J, \neg K \vdash \omega \quad (J, \neg K = J \cup (\neg K)).$$

By Proposition 19, this definition is indeed a generalization of the preceding one.

Remark *A priori*, the definition would seem to depend on the provable sentence θ. However, it is easy to see that any other provable sentence θ' yields an equivalent definition, since

$$J \vdash \neg\theta \quad \text{if and only if} \quad J \vdash \neg\theta'.$$

In fact:

$$\theta \vdash \theta' \quad \text{and} \quad \theta' \vdash \theta, \quad \text{thus} \quad \neg\theta' \vdash \neg\theta \quad \text{and} \quad \neg\theta \vdash \neg\theta'.$$

We can also replace $\omega = \neg\theta$ by any sentence ω' such that $\neg\omega'$ is provable, for then

$$\theta \vdash \neg\omega' \quad \text{and} \quad \neg\omega' \vdash \theta;$$

therefore $\neg\neg\omega' \vdash \omega$ and $\omega \vdash \neg\neg\omega'$, so that $\omega' \vdash \omega$ and $\omega \vdash \omega'$.

Remarks $J \vdash K$ does not mean that $K \subset T(J)$, i.e. $J \vdash B$ for every $B \in K$; however:

If $K \subset T(J)$ and $K \neq \emptyset$, then $J \vdash K$, since for any $B \in K$:

$$J \vdash B, \quad \text{therefore } J, \neg B \vdash \omega \text{ and } \textit{a fortiori } J, \neg K \vdash \omega.$$

In fact, this proves an even stronger assertion:

If there exists $B \in K$ such that $J \vdash B$, then $J \vdash K$.

The converse is false; for example, let $J = \emptyset$, $K = \{\omega, \theta\}$. Then $J, \neg K = \{\neg\omega, \omega\}$, thus $J, \neg K \vdash \omega$ and $J \vdash K$. But $K \not\subset T(\emptyset) = T$.

It may also be seen that $J \vdash K$ does not necessarily imply the existence of a sentence $B \in K$ such that $J \vdash B$; for example:

Let $J = \emptyset$ and $K = \{u^0, \neg u^0\}$, $u^0 \notin T(J)$ and $\neg u^0 \notin T(J)$. But $\neg u^0, \neg\neg u^0 \vdash \omega$, in fact, $\neg u^0 \vdash \neg u^0$.

Note that $J \vdash \emptyset$ means $J \vdash \omega$; we shall interpret this fact in the second part of this chapter.

If $J \neq \emptyset$ then always $J \vdash J$ (but $\emptyset \vdash \emptyset$ is false).

Global deducibility also has a number of interesting properties:

20 Proposition *$J \vdash K$ if and only if there exist a finite subset J^0 of J and a finite subset K^0 of K such that $J^0 \vdash K^0$.*

Proof $J \vdash K$ if and only if $J, \neg K \vdash \omega$, which is true if and only if there exists a finite subset $J^0, \neg K^0$ of $J, \neg K$ such that $J^0, \neg K^0 \vdash \omega$.

21 Proposition *$J \vdash A, B$ if and only if $J \vdash A \vee B$ (or $J \vdash B \vee A$).*

Proof The following assertions are equivalent:

$J \vdash A \vee B$

$J, \neg A \vdash B$ (Proposition 14)

$J, \neg A, \neg B \vdash \omega$ (Proposition 19)

$J \vdash A, B$.

Corollary *Let K be a finite nonempty set of sentences $\{B_1, \ldots, B_n\}$, and set*

$$D_1 = B_1$$
$$D_2 = D_1 \vee B_2$$
$$\ldots$$
$$D_n = D_{n-1} \vee B_n.$$

Then $J \vdash K$ if and only if $J \vdash D_n$ (for any J).

Proof By induction on n.

For $n = 1$ the assertion is trivial.

Assume the assertion true for $n - 1$. Then the following assertions are equivalent.

$J \vdash K$

$J, \neg K \vdash \omega$

$J, \neg B_n \vdash B_1, \ldots, B_{n-1}$

$J, \neg B_n \vdash D_{n-1}$ (by the induction hypothesis)

$J, \neg B_n, \neg D_{n-1} \vdash \omega$

$J \vdash B_n, D_{n-1}$

$J \vdash D_n$ (Proposition 21)

Note that the order of the sentences B_i is immaterial. We can therefore denote the disjunction of the sentences of K, in any order, by $\vee K$. Then, if K is finite, $J \vdash K$ if and only if $J \vdash \vee K$.

In particular, if both J and K are finite, then $J \vdash K$ if and only if $\wedge J \vdash \vee K$.

In the general case, for any J and K, $J \vdash K$ if and only if there exist a finite subset J^0 of J and a finite subset K^0 of K such that $\wedge J^0 \vdash \vee K^0$.

Remark Global deducibility is not transitive: if $J \vdash K$ and $K \vdash K'$ we cannot conclude that $J \vdash K'$.

Example $\varnothing \vdash u^0, \neg u^0$ (as we have already seen)

$$u^0, \neg u^0 \vdash \omega \quad \text{since} \quad u^0 \vdash u^0.$$

But $\varnothing \vdash \omega$ is false, since $\omega \notin T$.

However, we can prove the following particular case:

If $J \vdash A$ and $A \vdash K$, then $J \vdash K$.

In fact, $J \vdash A$ and $A \vdash \vee K^0$, where K^0 is a certain finite subset of K. Therefore $J \vdash \vee K^0$ (by Proposition 8). Thus $J \vdash K^0$ and *a fortiori* $J \vdash K$.

22 PROPOSITION *$J \vdash K, K'$ if and only if $J, \neg K' \vdash K$;
$J, J' \vdash K$ if and only if $J \vdash \neg J', K$.*

Proof The first equivalence is a trivial result of the definition.

The second is less evident. We first prove that $J, A \vdash B$ if and only if $J \vdash \neg A, B$. This follows from the following sequence of equivalent statements:

$$J, A \vdash B$$
$$J \vdash A \to B$$
$$J \vdash \neg B \to \neg A \quad \text{(by Proposition 17 and S7)}$$
$$J, \neg B \vdash \neg A$$
$$J \vdash \neg A, B.$$

We now claim that $J, J' \vdash \omega$ if and only if $J \vdash \neg J'$, provided J' is finite. This is proved by induction on the number n of sentences in J':

If $n = 0$ this is trivial.

Assume the assertion true for $n-1$, and let $J' = \{B_1, \ldots, B_n\}$ and $J'' = \{B_1, \ldots, B_{n-1}\}$.

$J, J' \vdash \omega$ if and only if
$$J, J'', B_n \vdash \omega$$
$$J, B_n \vdash \neg J'' \quad \text{by the induction hypothesis}$$
$$J, B_n \vdash \vee \neg J''$$
$$J \vdash \neg B_n, \vee \neg J'' \quad \text{by previous arguments}$$
$$J, \neg\neg B_n \vdash \vee \neg J''$$
$$J, \neg\neg B_n \vdash \neg J''$$
$$J \vdash \neg B_n, \neg J''$$
$$J \vdash \neg J'.$$

Finally, for arbitrary sets J, J', K:

If $J, J' \vdash K$, then there exist finite subsets J^0, J'^0, K^0 of J, J', K such that $J^0, J'^0 \vdash K^0$. Therefore $J^0, J'^0, \neg K^0 \vdash \omega$, whence $J^0, \neg K^0 \vdash \neg J'^0$ and $J^0 \vdash \neg J'^0, K^0$. *A fortiori* $J \vdash \neg J', K$; and conversely.

Note in particular: $J \vdash K$ if and only if $J, \neg K \vdash \emptyset$ or $\neg K \vdash \neg J$.

The following properties concern only the predicate calculus. For any set of sentences J, I_J (the basis of J) is the set of individuals which have at least one occurrence in at least one sentence of J.

23 PROPOSITION *For any J, K and individual a:*

$$\frac{J \vdash K, (a/x)f^x}{J \vdash K, \exists x\,[f^x]} \quad \text{and} \quad \frac{J, (a/x)f^x \vdash K}{J, \forall x\,[f^x] \vdash K}.$$

Proof $J \vdash K, (a/x)f^x$

$J, \neg K \vdash (a/x)f^x$; but $(a/x)f^x \vdash \exists x\,[f^x]$, therefore

$J, \neg K \vdash \exists x\,[f^x]$

$J \vdash K, \exists x\,[f^x]$.

The proof of the second assertion is analogous:

$J, (a/x)f^x \vdash K$

$J, \neg K \vdash \neg(a/x)f^x$; but $\forall x\,[f^x] \vdash (a/x)f^x$, therefore

$\neg(a/x)f^x \vdash \neg\forall x\,[f^x]$

$J, \neg K \vdash \neg\forall x\,[f^x]$

$J, \forall x\,[f^x] \vdash K$.

24 PROPOSITION
$$\frac{J, (a/x)f^x \vdash K}{J, \exists x\,[f^x] \vdash K} \quad \text{and} \quad \frac{J \vdash K, (a/x)f^x}{J \vdash K, \forall x\,[f^x]}.$$

Proof If $J, (a/x)f^x \vdash K$, then there exist finite subsets J^0, K^0 of J, K such that $J^0, (a/x)f^x \vdash K^0$. Then:

$(a/x)f^x \vdash \neg J^0, K^0$

$(a/x)f^x \vdash \vee(\neg J^0, K^0)$

$\emptyset \vdash (a/x)f^x \to \vee(\neg J^0, K^0)$

$\emptyset \vdash \exists x\,[f^x] \to \vee(\neg J^0, K^0)$ (rule of inference)

$\exists x\,[f^x] \vdash \vee(\neg J^0, K^0)$

$\exists x\,[f^x] \vdash \neg J^0, K^0$

$\exists x\,[f^x] \vdash \neg J, K$

$J, \exists x\,[f^x] \vdash K$.

Similarly: $J \vdash K, (a/x) f^x$

$J^0 \vdash K^0, (a/x) f^x$

$\wedge(J^0, \neg K^0) \vdash (a/x) f^x$

$\emptyset \vdash \wedge(J^0, \neg K^0) \rightarrow (a/x) f^x$

$\emptyset \vdash \wedge(J^0, K^0) \rightarrow \forall x\, [f^x]$ (rule of inference)

etc.

Particular cases If $a \notin I_{J, f^x}$, then $\dfrac{J \vdash (a/x) f^x}{J \vdash \forall x\, [f^x]}$ ("Principle of generalization").

If $a \notin I_{K, f^x}$, then $\dfrac{(a/x) f^x \vdash K}{\exists x\, [f^x] \vdash K}$.

25 Proposition *If* $a \notin I_{J, f^x}$, *then* $\dfrac{J \vdash \exists x\, [f^x],\ (a/x) f^x \vdash K}{J \vdash K}$.

In fact, $(a/x) f^x \vdash K$ implies $\exists x\, [f^x] \vdash K$.

26 Proposition $\forall x\, [f^x] \vdash \neg \exists x\, [\neg f^x]$ *and* $\neg \exists x\, [\neg f^x] \vdash \forall x\, [f^x]$.

Proof For any a, $\forall x\, [f^x] \vdash (a/x) f^x$

$\neg (a/x) f^x \vdash \forall x\, [f^x]$

$(a/x) \neg f^x \vdash \neg \forall x\, [f^x]$, in particular, for $a \notin I_{f^x}$

$\exists x\, [\neg f^x] \vdash \forall x\, [f^x]$

$\forall x\, [f^x] \vdash \neg \exists x\, [\neg f^x]$.

Similarly: $(a/x) \neg f^x \vdash \exists x\, [\neg f^x]$

$\neg \exists x\, [\neg f^x] \vdash (a/x) f^x$, and so for $a \notin I_{f^x}$:

$\neg \exists x\, [\neg f^x] \vdash \forall x\, [f^x]$.

Remark 1 The following assertions are equivalent:

(1) $\forall x\, [f^x] \in T$,

(2) for all a, $(a/x) f^x \in T$,

(3) there exists $a \notin I_{f^x}$ such that $(a/x) f^x \in T$.

Proof $1 \Rightarrow 2$: $\emptyset \vdash \forall x\, [f^x]$ and $\forall x\, [f^x] \vdash (a/x) f^x$; therefore $\emptyset \vdash (a/x) f^x$.
$2 \Rightarrow 3$: trivial.
$3 \Rightarrow 1$: if $\emptyset \vdash (a/x) f^x$, where $a \notin I_{f^x}$, apply the principle of generalization.

On the other hand, it is important to note that $\exists x\, [f^x] \in T$ does *not* imply that there exists a such that $(a/x) f^x \in T$.

To see this, consider the following example:

$A = \exists x\, [r_0^1 x \vee \forall y\, [\neg r_0^1 y]] = \exists x\, [f^x]$, where $f^x = r_0^1 x \vee \forall y\, [\neg r_0^1 y]$.

If a is an individual, then $(a/x)f^x = r_0^1 a \vee \forall y\,[\neg r_0^1 y]$. Let h be a system of truth values such that

$$h\,(r_0^1 a) = 0,$$

$$h\,(r_0^1 b) = 1 \quad \text{if } b \neq a.$$

Then $\tilde{h}\,((a/x)f^x) = 0$, so that $(a/x)f^x \notin T$ for any individual a. On the other hand, $A \in T$, as we shall now prove.

$$r_0^1 a \vdash \exists x\,[r_0^1 x]$$
$$\neg\exists x\,[r_0^1 x] \vdash \neg r_0^1 a$$
$$\neg\exists x\,[r_0^1 x] \vdash \forall y\,[\neg r_0^1 y] \quad \text{(principle of generalization)}$$

so that $\emptyset \vdash \exists x\,[r_0^1 x] \vee \forall y\,[\neg r_0^1 y]$.

Now $r_0^1 a \vdash r_0^1 a \vee \forall y\,[\neg r_0^1 y]$

therefore $r_0^1 a \vdash \exists x\,[r_0^1 x \vee \forall y\,[\neg r_0^1 y]]$

hence $\exists x\,[r_0^1 x] \vdash \exists x\,[r_0^1 x \vee \forall y\,[\neg r_0^1 y]]$

and $\forall y\,[\neg r_0^1 y] \vdash r_0^1 a \vee \forall y\,[\neg r_0^1 y]$

so that $\forall y\,[\neg r_0^1 y] \vdash \exists x\,[r_0^1 x \vee \forall y\,[\neg r_0^1 y]]$.

Finally:

$$\exists x\,[r_0^1 x] \vee \forall y\,[\neg r_0^1 y]] \vdash \neg\exists x\,[r_0^1 x] \to \forall y\,[\neg r_0^1 y]$$
$$\exists x\,[r_0^1 x] \vee \forall x\,[\neg r_0^1 y],\ \neg\exists x\,[r_0^1 x] \vdash \forall y\,[\neg r_0^1 y]$$
$$\exists x\,[r_0^1 x] \vee \forall y\,[\neg r_0^1 y],\ \neg\exists x\,[r_0^1 x] \vdash \exists x\,[r_0^1 x \vee \forall y\,[\neg r_0^1 y]]$$
$$\exists x\,[r_0^1 x] \vee \forall y\,[\neg r_0^1 y],\ \neg\exists x\,[r_0^1 x \vee \forall y\,[\neg r_0^1 y]] \vdash \exists x\,[r_0^1 x]$$
$$\exists x\,[r_0^1 x] \vee \forall y\,[\neg r_0^1 y],\ \neg\exists x\,[r_0^1 x \vee \forall y\,[\neg r_0^1 y]] \vdash \exists x\,[r_0^1 x \vee \forall y\,[\neg r_0^1 y]].$$

Thus $\exists x\,[r_0^1 x] \vee \forall y\,[\neg r_0^1 y] \vdash A$, which implies that $A \in T$.

It is easily seen that A is universally valid: Let h be any system of truth values. Then, if there exists a such that $h\,(r_0^1 a) = 1$, then $\tilde{h}\,((a/x)f^x) = 1$; otherwise, $h\,(r_0^1 a) = 0$ for all a, so that $\tilde{h}\,(\forall y\,[\neg r_0^1 y]) = 1$. In either case $\tilde{h}(A) = 1$.

However, since the completeness theorem is not yet at our disposal we cannot infer that A is a provable sentence.

Remark 2 In the predicate calculus, the relation of deducibility is usually defined in a different way and limited to finite sets of hypotheses J. A deduction is any finite sequence of sentences $A_1, \ldots, A_n$ each of which satisfies at least one of the following five conditions:

(1) $A_i \in J$;

(2) A_i is an instance of an axiom schema;

(3) there exist $j < i$ and $h < i$ such that $A_h = A_j \to A_i$;

(4) $A_i = \exists x\,[f^x] \to B$ and there exists $j < i$ such that $A_j = (a/x) f^x \to B$, where $a \notin I_{J,B,f^x}$;

(5) $A_i = B \to \forall x\,[f^x]$ and there exists $j < i$ such that $A_j = B \to (a/x) f^x$, where $a \notin I_{J,B,f^x}$.

Let $J \vdash^* A$ mean that there exists a deduction of A satisfying this definition. We claim that when J is finite the relations $J \vdash A$ and $J \vdash^* A$ are equivalent. In fact:

If $A_1, \ldots, A_n$ is a deduction in the sense $J \vdash A$, it suffices to replace every provable sentence A_i of the type $A_i = \exists x\,[f^x] \to B$ or $A_i = B \to \forall x\,[f^x]$ by a formal proof of this sentence which contains no individuals other than those in the infinite set $I_{A_i} \cup \complement I_J$ (this is possible because of Proposition 8 of Chapter I). The result is a deduction in the sense $J \vdash^* A$.

Conversely, if $A_1, \ldots, A_n$ is a deduction in the sense $J \vdash^* A$, we shall prove by induction on n that $J \vdash A$.

If $A \in J$ or A is an instance of an axiom schema, this is trivial.

If A is obtained by modus ponens, the assertion is again trivial.

If $A = \exists x\,[f^x] \to B$ is obtained from $(a/x) f^x \to B$, where $a \notin I_{J,B,f^x}$, then by the induction hypothesis $J \vdash (a/x) f^x \to B$. Thus, $J, (a/x) f^x \vdash B$, whence $J, \exists x\,[f^x] \vdash B$, so that $J \vdash A$.

Note that the restriction $a \notin I_J$ in conditions (4) and (5) is essential, as the following example shows.

Let J consist of the single sentence $r_0^1 a \to r_0^2 bc$, i.e., $(a/x) f^x \to B$, where $f^x = r_0^1 x$, $B = r_0^2 bc$. Then $a \notin I_{B,f^x}$, but it is not true that $J \vdash \exists x\,[r_0^1 x] \to r_0^2 bc$; for $(r_0^1 a \to r_0^2 bc) \to (\exists x\,[r_0^1 x] \to r_0^2 bc) \notin T$, as may be seen by considering a system of truth values h such that

$$h(r_0^2 bc) = 0, \quad h(r_0^1 a) = 0, \quad h(r_0^1 b) = 1.$$

2 Deductive systems

Consistent subsets

A subset J of E is said to be *consistent* if $T(J) \neq E$; otherwise it is inconsistent.

Examples E is inconsistent.

T is consistent, since $T(T) = T \neq E$—this is one of the consistency theorems.

Any subset J of T is consistent, since $T(J) = \mathrm{T}$. In general, any subset of a consistent subset is consistent.

Note the following properties.

27 PROPOSITION *The following assertions are equivalent:*

(1) J is inconsistent;

(2) there exists a sentence A such that $J \vdash A$ and $J \vdash \neg A$;

(3) $J \vdash \omega$ (or $J \vdash \emptyset$).

Proof If J is inconsistent, the other two assertions are trivially true.

If there exists a sentence A such that $J \vdash A$ and $J \vdash \neg A$, let B be any sentence. Then B is deducible from J as follows (cf. consistency):

$$\begin{array}{l}
\ldots, A \quad \text{(deduction of } A \text{ from } J) \\
\neg A \\
\neg A \rightarrow (\neg B \rightarrow \neg A) \\
\neg B \rightarrow \neg A \\
(\neg B \rightarrow \neg A) \rightarrow (A \rightarrow B) \\
A \rightarrow B \\
B.
\end{array}$$

Thus $J \vdash B$ for all B, so that J is inconsistent. In particular, $J \vdash \omega$.

If $J \vdash \omega$, i.e. $J \vdash \neg\theta$, then (since $\theta \in T$) it is also true that $J \vdash \theta$, and this implies the preceding property.

28 PROPOSITION *The following assertions are equivalent:*

(1) J is inconsistent;

(2) there exists a finite inconsistent subset J^0 of J.

To see this it is sufficient to recall that J is inconsistent if and only if $J \vdash \omega$.

29 PROPOSITION *For any set J, $J \vdash K$ if and only if $J, \neg K$ is inconsistent, since by definition $J \vdash K$ if and only if $J, \neg K \vdash \omega$.*

30 PROPOSITION *For the case of a single sentence A, A is inconsistent if and only if $\neg A \in T$.*

For $A \vdash \omega$ if and only if $\emptyset \vdash \neg A$.

31 PROPOSITION *If J is a consistent set, then, for any sentence A, at least one of the sets J, A and $J, \neg A$ is consistent.*

In fact, were both sets inconsistent, we would have

$$J, A \vdash \omega, \qquad \text{therefore } J \vdash \neg A;$$

$$J, \neg A \vdash \omega, \quad \text{therefore } J \vdash A.$$

But this contradicts the consistency of J.

Note that both these sets may be consistent.

Deductive systems—complete deductive systems

A deductive system is any *consistent* subset Δ of E such that

$$T(\Delta) = \Delta.$$

Example T is a deductive system (the smallest possible, in the sense of set inclusion). In general, if J is consistent then $T(J)$ is a deductive system, since $T(T(J)) = T(J)$.

A deductive system may be characterized as any subset of E with the following properties:

(1) $T \subset \Delta \subsetneqq E$;
(2) Δ is closed under modus ponens: if $A \in \Delta$ and $A \to B \in \Delta$, then $B \in \Delta$.

In fact:
Any deductive system clearly has these properties.
Conversely, if Δ is a subset with these properties, it is easy to see that if $\Delta \vdash A$ then $A \in \Delta$ (by induction on the length of a deduction of A); thus $T(\Delta) = \Delta \neq E$.
One particular case, which plays a fundamental role in the sequel, is that of a *complete deductive system* (abbreviation: cds), which is defined as any subset V of E such that

(1) V is a deductive system;
(2) for any sentence A, either $A \in V$ or $\neg A \in V$.

Note that T (as a deductive system) is not complete, since for any elementary sentence u we have both $u \notin T$ and $\neg u \notin T$.
The existence of complete deductive systems will be proved later.

32 Proposition *Any deductive system is the union of equivalence classes modulo a certain relation R.*

Proof Let R be the relation: $A \vdash B$ and $B \vdash A$ (or $A \leftrightarrow B \in T$). Then if $A \in \Delta$ and B is such that ARB, we have $\Delta \vdash A$ and $A \vdash B$, therefore $\Delta \vdash B$, and so $B \in \Delta$.

33 Proposition *The set of deductive systems, ordered by set inclusion, is inductive.*

Proof Let $\{\Delta_i\}$ be a linearly ordered family of deductive systems, and set $\Delta = \cup\Delta_i$. We claim that Δ is also a deductive system (which is thus the supremum of the family).

$T \subset \Delta_i$ for all i, therefore $T \subset \Delta$.

Let $A \in \Delta$ and $A \to B \in \Delta$. Then there exist i_1 such that $A \in \Delta_{i_1}$ and i_2 such that $A \to B \in \Delta_{i_2}$. Now Δ_{i_1} and Δ_{i_2} are comparable, say $\Delta_{i_1} \subset \Delta_{i_2}$, so that $B \in \Delta_{i_2}$ and *a fortiori* $B \in \Delta$.

Finally, Δ is consistent, since otherwise $\omega \in \Delta$ and thus $\omega \in \Delta_i$ for some i.

By Zorn's Lemma, Proposition 33 implies the existence of maximal deductive systems. More precisely: every deductive system is contained in a maximal deductive system.

34 PROPOSITION *Every maximal deductive system is a cds, and* vice versa.

Proof Let V be a maximal deductive system, and A any sentence. Suppose that $A \notin V$. Then $V \vdash A$ is false, and so (Proposition 29) $V, \neg A$ is consistent. Therefore $T(V, \neg A)$ is a deductive system. But $V = T(V) \subset T(V, \neg A)$ and V is maximal, therefore $V = T(V, \neg A)$. Thus $\neg A \in V$ and V is complete.

Conversely, let V be a cds, and suppose that $V \subset \Delta$, where Δ is a deductive system. Let $A \in \Delta$. If $\neg A \in V$, this implies $\neg A \in \Delta$, which is impossible; therefore $A \in V$. Thus $\Delta \subset V$ and so $\Delta = V$ and V is maximal.

We have thus proved the existence of cds. Later we shall prove the existence of a certain class of cds without using the axiom of choice [used here via Zorn's Lemma].

Another version of this result is the following:

THEOREM OF LINDENBAUM-TARSKI *Every deductive system possesses a complete extension.*

In other words, every deductive system is contained in a cds.

35 PROPOSITION *A deductive system is the intersection of the cds that contain it.*

Proof Let Δ be a deductive system, and $\mathscr{V}_\Delta$ the family of cds V such that $V \supset \Delta$. Then it is trivial that $\Delta \subset \bigcup_{\mathscr{V}_\Delta} V$.

Conversely, let A be a sentence belonging to all cds V containing Δ, and suppose that $A \notin \Delta$. Then $\Delta \vdash A$ is false, and therefore $\Delta, \neg A$ is consistent.

Thus $T(\Delta, \neg A)$ is a deductive system with a complete extension V. Then

$$\Delta = T(\Delta) \subset T(\Delta, \neg A) \subset V, \quad \text{so that } V \in \mathscr{V}_\Delta.$$

Therefore $A \in V$; but $\neg A \in V$, and this is a contradiction. Therefore $A \in \Delta$.

COROLLARY *T is the intersection of all cds, since T is contained in every cds.*

36 PROPOSITION *If V is a cds, then:*

(a) $\neg A \in V$ *if and only if* $A \notin V$.
(b) $A \wedge B \in V$ *if and only if* $A \in V$ *and* $B \in V$.
(c) $A \vee B \in V$ *if and only if* $A \in V$ *or* $B \in V$.
(d) $A \to B \in V$ *if and only if* $A \notin V$ *or* $B \in V$.
(e) $A \leftrightarrow B \in V$ *if and only if either* $A \in V$ *and* $B \in V$ *or* $A \notin V$ *and* $B \notin V$.

For the predicate calculus L″:

(f_1) *If there exists a such that* $(a/x)f^x \in V$, *then* $\exists x\,[f^x] \in V$.
(g_1) *If* $\forall x\,[f^x] \in V$, *then, for any a,* $(a/x)f^x \in V$.

Proof (a) If $A \notin V$, then $\neg A \in V$, since V is complete.* If $\neg A \in V$, then $A \notin V$, since V is consistent.*

(b) will be proved after (c).

(c) If $A \vee B \in V$, then $V \vdash A \vee B$, therefore $V, \neg A \vdash B$. Suppose that $A \notin V$; then $\neg A \in V$, therefore $V \vdash B$ and so $B \in V$.

If $A \in V$ or $B \in V$, then $V \vdash A$ or $V \vdash B$, *a fortiori* $V \vdash A, B$, and therefore $V \vdash A \vee B$.

(b) If $A \wedge B \in V$, then $V \vdash A \wedge B$, therefore $V \vdash \neg(\neg A \vee \neg B)$, so that $\neg(\neg A \vee \neg B) \in V$ or $\neg A \vee \neg B \notin V$. Consequently, $\neg A \notin V$ and $\neg B \notin V$, and therefore $A \in V$ and $B \in V$.

If $A \in V$ and $B \in V$, reverse the preceding argument.

(d) $V \vdash A \to B$ is equivalent to $V, A \vdash B$, hence to $V \vdash \neg A, B$, and finally to $V \vdash \neg A \vee B$.

(e) $V \vdash A \leftrightarrow B$ if and only if $V \vdash A \to B$ and $V \vdash B \to A$.

(f_1) If $V \vdash (a/x)f^x$, then since $(a/x)f^x \vdash \exists x\,[f^x]$ we have $V \vdash \exists x\,[f^x]$.

(g_1) If $V \vdash \forall x\,[f^x]$, then since $\forall x\,[f^x] \vdash (a/x)f^x$, we have $V \vdash (a/x)f^x$ for any a.

---

* *Translator's note* The words "complete" and "consistent" are interchanged in the original French text.

Truth sets

Let us return to the concept of a system of truth values.

Let h be any system of truth values, and denote the set of h-true sentences by V_h; in other words,

$$V_h = \tilde{h}^{-1}(1)$$

($\tilde{h}$ is the characteristic function of V_h).

It is easy to see that V_h is always a cds:

For any $A \in T$ we have $\tilde{h}(A) = 1$ (Consistency Theorem); therefore $T \subset V_h$.

If $A \in T$ then $\tilde{h}(\neg A) = 0$; therefore $\neg A \notin V_h$ and so $V_h \neq E$.

Let $A \in V_h$ and $A \to B \in V_h$; then $\tilde{h}(A) = 1$ and $\tilde{h}(A \to B) = \tilde{h}(A) \to \tilde{h}(B) = 1$, so that $\tilde{h}(B) = 1$ and $B \in V_h$.

Finally, for any sentence A either $\tilde{h}(A) = 1$, and then $A \in V_h$, or $\tilde{h}(A) = 0$ so that $\tilde{h}(\neg A) = 1$ and $\neg A \in V_h$.

Thus V_h is indeed a cds. We have thus proved the existence of a certain class of complete deductive systems.

These special complete deductive systems play an extremely important role. We call them *truth sets.**

A truth set is thus completely determined by h (which may be any mapping of $\mathfrak{a}$ into $\mathbb{U}$) or, equivalently, by its intersection with $\mathfrak{a}$: $V_h \cap \mathfrak{a}$ (whose characteristic function is h).

The intersection of all truth sets is the set of universally true sentences.

Any truth set clearly satisfies properties a, b, c, d, e, f_1, g_1 of Proposition 36; however, for the calculus L'' we can state a more precise result: properties f_1 and g_1 may be replaced by the following:

(f) $\exists x\,[f^x]$ if and only if there exists an individual a such that $(a/x)f^x \in V_h$;

(g) $\forall x\,[f^x]$ if and only if for any individual a we have $(a/x)f^x \in V_h$.

Proof If $\exists x\,[f^x] \in V_h$, then $\tilde{h}\,(\exists x\,[f^x]) = \sup_a \tilde{h}\,((a/x)f^x) = 1$, therefore there exists a such that $\tilde{h}\,((a/x)f^x) = 1$.

If $(a/x)f^x \in V_h$ for any a, then $\tilde{h}\,((a/x)f^x) = 1$ for any a, so that $\inf_a \tilde{h}\,((a/x)f^x) = \tilde{h}\,(\forall x\,[f^x]) = 1$ and therefore $\forall x\,[f^x] \in V_h$.

Truth sets may also be characterized without recourse to systems of truth values:

---

* *Translator's note* French *validation*, which has no direct equivalent in English. Our term *truth set* follows R. Smullyan *First-Order Logic*, Springer-Verlag, Berlin (1968).

37 Proposition *If V is a subset of E which satisfies the conditions*

a, b, c, d, e *for the calculus L′,*

a, b, c, d, e, f, g *for the calculus L″,*

then V is a truth set.

Proof Let H be the characteristic function of V. Then the above properties are equivalent to the following:

$$\begin{aligned} H(\neg A) &= \neg H(A) \\ H(A \wedge B) &= H(A) \wedge H(B) \\ H(A \vee B) &= H(A) \vee H(B) \\ H(A \rightarrow B) &= H(A) \rightarrow H(B) \\ H(A \leftrightarrow B) &= H(A) \leftrightarrow H(B) \\ H(\exists x\, [f^x]) &= \sup_a H((a/x)f^x) \\ H(\forall x\, [f^x]) &= \inf_a H((a/x)f^x). \end{aligned}$$

By the corollary to Proposition 1 of Chapter II, H is completely determined by its restriction h to $\mathfrak{a}$, and therefore $V = V_h$. This yields a very important theorem:

38 Proposition *For the calculus L′, any complete deductive system is a truth set, and* vice versa.

No such result is available for the calculus L'', since in general a cds does not satisfy conditions f and g (only f_1 and g_1). Here lies the fundamental difference between the two logical calculi; essentially, this is the root of the difficulties involved in proving completeness for the predicate calculus.

To be precise, we prove that in the calculus L'' there exist cds which are not truth sets.

Let I^+ be the set of individuals consisting of I and one new symbol ε, and let E^+ be the superlogic of E based on I^+. We know that $T = T^+ \cap E$.

Consider a predicate r_n^p of nonzero weight and the quantifiable formula $r_n^p\, xa_2 \cdots a_p$ (where $a_2, \ldots, a_p \in I$). Let h^+ be a system of truth values in E^+ such that

$$h^+(r_n^p\, a_1 a_2 \cdots a_p) = 0 \quad \text{for all } a_1 \in I;$$

$$h^+(r_n^p\, \varepsilon\, a_2 \cdots a_p) = 1.$$

Then $\tilde{h}^+(\exists x\, [f^x]) = 1$.

Let V be the set of sentences $A \in E$ such that $\tilde{h}^+(A) = 1$.

V is a cds in E:

If $A \in T$, then $A \in T^+$, therefore $\tilde{h}^+(A) = 1$, and so $T \subset V$.

If $\tilde{h}^+(A \to B) = 1$, then $\tilde{h}^+(B) = 1$.

$V \neq E$, since $r_n^p a_1 a_2 \cdots a_p \notin V$ for any $a_1 \in I$.

If $A \notin V$, then $\tilde{h}^+(A) = 0$, thus $\tilde{h}^+(\neg A) = 1$, therefore $\neg A \in V$.

However, V is not a truth set in E: $\exists x\,[f^x] \in V$, but $(a/x)f^x \notin V$ for all $a \in I$.

Inverse image of a cds under a substitution

39 PROPOSITION *Let s be a simultaneous substitution of individuals and I^* an infinite subset of I such that $s(I) \subset I^*$. Then if V^* is a cds in the sublogic E^* based on I^*, then $s^{-1}V^*$ is a cds in E. ($s^{-1}V^*$ denotes the set of sentences $A \in E$ such that $s(A) \in V^*$.)*

Proof If $A \in T$, then $s(A) \in T \cap E(I^*)$. Therefore $s(A) \in T^*$, and so $s(A) \in V^*$. This shows that $T \subset s^{-1}V^*$.

If $s(A) \in V^*$ and $s(A \to B) \in V^*$, then $s(B) \in V^*$.

If $A \in s^{-1}V^*$, then $\neg A \notin s^{-1}V^*$, therefore $s^{-1}V^* \neq E$.

If $A \notin s^{-1}V^*$, then $s(A) \notin V^*$, and so $s(\neg A) \in V^*$.

Remark In particular:

For any cds V in E and any simultaneous substitution s, $s^{-1}V$ is a cds in E.

Note that this is false for the direct image sV.

For truth sets, we have the following result, which is somewhat weaker:

40 PROPOSITION *If I^* is an infinite subset of I and s is a mapping of I* onto *I^*, then, for any truth set V^* of E^*, $s^{-1}V^*$ is a truth set of E.*

Proof We already know that $s^{-1}V^*$ is a cds. It remains to prove properties f and g.

Let $\exists x\,[f^x] \in s^{-1}V^*$. Then $s(\exists x\,[f^x]) = \exists x\,[s(f^x)] \in V^*$, so that there exists $a \in I^*$ such that $(a/x)\,s(f^x) \in V^*$. But s maps I onto I^*, therefore there exists $a' \in I$ such that $a = s(a')$. Then $(a/x)\,s(f^x) = s\,(a'/x)f^x$, whence $(a'/x)f^x \in s^{-1}V^*$.

Similarly, if $(a/x)f^x \in s^{-1}V^*$ for all $a \in I$, then $\forall x\,[f^x] \in s^{-1}V^*$.

Note that if $\tilde{h}^*$ is the characteristic function of V^* in E^*, then the characteristic function of $s^{-1}V^*$ in E is $\tilde{h}^*_{\mathrm{o}}s$.

CHAPTER IV

Completeness of the propositional calculus

The results proved at the end of the previous chapter yield an immediate proof of the following theorem, for the calculus L':

1 First (semantic) completeness theorem

Every universally valid sentence is provable.

Proof If $\tilde{h}(A) = 1$ for any system of truth values h, then $A \in V_h$ for any truth set, i.e. for any cds. But T is the intersection of all cds, therefore $A \in T$.

This theorem is the converse of the First (Semantic) Consistency Theorem. In view of these two theorems we can now assert that the following two statements are equivalent:

A is a provable sentence.

$\tilde{h}(A) = 1$ for any system of truth values.

We thus have a *decision procedure* for the propositional calculus, that is to say, a systematic method by which we can determine, in a finite number of steps, whether a given sentence is provable or not:

Given a sentence A, calculate all the truth values $\tilde{h}(A)$ of A; only the values of h for propositional variables occurring in A participate in the calculation.

In general, let $\mathfrak{a}_A$ denote the set of propositional variables having at least one occurrence in A. If $\mathfrak{a}_A$ consists of n elements, only 2^n mappings need be checked in the decision procedure for A.

Example $A = u \to u$.

If $h(u) = 1$, then $\tilde{h}(A) = 1 \to 1 = 1$;

if $h(u) = 0$, then $\tilde{h}(A) = 0 \to 0 = 1$.

Thus $A \in T$.

(2) $A = u \to u \wedge v$.

If	$h(u) = 1$ and	$h(v) = 1$:	$\tilde{h}(A) = 1 \to 1 = 1$;
	1	0	$1 \to 0 = 0$;
	0	1	$0 \to 0 = 1$;
	0	0	$0 \to 0 = 1$.

Thus $A \notin T$.

This procedure also yields certain additional information:

(1) It provides, so to speak, a truth-functional analysis of the sentence; e.g., in the preceding example:

A is h-false if and only if u is h-true and v is h-false.

(2) It enables us to find sentences that are equivalent (in the sense of the relation R introduced above) to the given sentence; in the preceding example: A is h-false if and only if $\neg u \vee v$ is h-false; thus $\bar{h}(A) = \bar{h}(\neg u \vee v)$ for all h, or $\bar{h}(A \leftrightarrow \neg u \vee v) = 1$ for all h. Thus $A \leftrightarrow \neg u \vee v \in T$. Therefore A is equivalent to the sentence $\neg u \vee v$.

We shall not go into the latter question here (it pertains to the study of normal forms).

The following scheme is convenient for the truth-functional analysis of a sentence.

Example $(u \wedge v) \vee (\neg u \wedge \neg w) \to (v \leftrightarrow w)$.

<table>
<tr><td colspan="4" align="center">u</td></tr>
<tr><td colspan="2" align="center">$(1 \wedge v) \vee (\neg 1 \wedge \neg w) \to (v \leftrightarrow w)$
$v \vee (0 \wedge \neg w) \to (v \leftrightarrow w)$
$v \vee 0 \to (v \leftrightarrow w)$
$v \to (v \leftrightarrow w)$</td><td colspan="2" align="center">$(0 \wedge w) \vee (\neg 0 \wedge \neg w) \to (v \leftrightarrow w)$
$0 \vee (1 \wedge \neg w) \to (v \leftrightarrow w)$
$0 \vee \neg w \to (v \leftrightarrow w)$
$\neg w \to (v \leftrightarrow w)$</td></tr>
<tr><td colspan="2" align="center">v</td><td colspan="2" align="center">w</td></tr>
<tr><td align="center">$1 \to (1 \leftrightarrow w)$
$1 \leftrightarrow w$

w
<table><tr><td>$1 \leftrightarrow 1$
1</td><td>$1 \leftrightarrow 0$
0</td></tr></table></td><td align="center">$0 \to (0 \leftrightarrow w)$
1</td><td align="center">$\neg 1 \to (v \leftrightarrow 1)$
$0 \to (v \leftrightarrow 1)$
1</td><td align="center">$\neg 0 \to (v \leftrightarrow 0)$
$1 \to (v \leftrightarrow 0)$
$v \leftrightarrow 0$
v
<table><tr><td>$1 \leftrightarrow 0$
0</td><td>$0 \leftrightarrow 0$
1</td></tr></table></td></tr>
</table>

The procedure is as follows. Identify $u, v, w, \ldots$ with elements of $\mathbb{U}$. Assign one of the propositional variables (in practice it is most convenient to consider that with the greatest number of occurrences) the value 1 on the left of the tableau and the value 0 on the right. Then carry out the calculations in $\mathbb{U}$, and assign values to another propositional variable, etc. The different possible truth values of the sentence A appear at the bottom of each column.

In the above example the sentence A is not provable; it is h-false in the following two cases:

u is h-true, v is h-true, w is h-false;
u is h-false, v is h-true, w is h-false.

In other words, A is h-false if and only if v is h-true and w is h-false, so that it is equivalent to $\neg v \vee w$.

We now proceed to other (syntactic) forms of the completeness theorem. To this end we use the concept of a multiple substitution instance of a sentence A (already studied in Chapter I):

A sentence B is said to be a multiple substitution instance of A if there exists a finite sequence of sentences $B_1 = A, B_2, \ldots, B_n = B$ such that for each sentence B_i $(2 \leqslant i \leqslant n)$ there exist $u_i \in \alpha$ and $C_i \in E$ such that $B_i = (C_i/u_i)\, B_{i-1}$.

We denote the set of multiple substitution instances of A by $\Sigma(A)$.

Remark If $A \in T$ then $\Sigma(A) \subset T$.

In fact, if $B \in \Sigma(A)$ and B is the last term of a sequence of substitutions

$$B_1 = A, B_2, \ldots, B_n = B,$$

it is immediate by induction that $B_i \in T$ for all i.

1 LEMMA *Let h be a system of truth values, A a sentence, and u a propositional variable. Let h' be the system of truth values defined by*

$$h'(u) = \tilde{h}(A),$$

$$h'(v) = h(v) \quad \textit{for } v \neq u.$$

Then for, every sentence B, $h'(B) = \tilde{h}\,((A/u)\,B)$

Proof By induction on the order n of B.

If $n = 0$, then B is an atomic sentence:

If $B = u$: $(A/u)\,B = A$ and indeed $\tilde{h}'(u) = h'(u) = \tilde{h}(A)$;

if $B = v \neq u$: $(A/u)\,B = B = v$ and $\tilde{h}'(B) = h'(v) = h(v) = \tilde{h}(v)$.

Now assume the assertion true for sentences of order up to n, and let B be of order $n+1$. Then we have the following possibilities:

(a) $B = \neg B'$:

$$\tilde{h}'(B) = \tilde{h}'(\neg B') = \neg \tilde{h}'(B') = \neg \tilde{h}\,((A/u)\,B') = \tilde{h}\,(\neg (A/u)\,B')$$
$$= \tilde{h}\,((A/u)\,\neg B') = \tilde{h}\,((A/u)\,B).$$

(b) $B = B' k B''$:

$$\tilde{h}'(B) = \tilde{h}'(B')\, k\, \tilde{h}'(B'') = \tilde{h}((A/u)B')k\tilde{h}((A/u)B'')$$
$$= \tilde{h}((A/u)B'k(A/u)B'')$$
$$= \tilde{h}((A/u)B).$$

The lemma is thus proved.

Let $v(A)$ denote the number of different propositional variables occurring in the sentence A (i.e. the number of elements in the set $\mathfrak{a}_A$).

2 Lemma *If $A \notin T$, there exists a sentence $B(A) \in \Sigma(A)$ such that*
a) if $v(A) > 1$ then $v(B(A)) = v(A) - 1$, $B(A) \notin T$;
b) if $v(A) = 1$ then $v(B(A)) = 1$, $\neg B(A) \in T$ (so that $B(A) \notin T$).

Proof Since $A \notin T$ there exists a system of truth values h such that $\tilde{h}(A) = 0$ (Semantic Completeness Theorem).

Choose $u \in \mathfrak{a}_A$, and consider the following cases.

(1) $v(A) > 1$ and $h(u) = 1$; put $A' = v \vee \neg v$, where $v \in \mathfrak{a}_A - \{u\}$.
(2) $v(A) > 1$ and $h(u) = 0$; put $A' = v \wedge \neg v$, where $v \in \mathfrak{a}_A - \{u\}$.
(3) $v(A) = 1$ and $h(u) = 1$; put $A' = u \vee \neg u$.
(4) $v(A) = 1$ and $h(u) = 0$; put $A' = u \wedge \neg u$.

In all cases we put $B(A) = (A'/u)\,A$. It is clear that $B(A) \in \Sigma(A)$.

First consider the case $v(A) > 1$. Then clearly $v(B(A)) = v(A) - 1$. Define h' by

$$h'(u) = \tilde{h}(A'),$$

$$h'(w) = h(w) \quad \text{for } w \neq u.$$

Then by Lemma 1 we have $\tilde{h}'(A) = \tilde{h}((A'/u)\,A) = \tilde{h}(B(A))$.
Now if $h(u) = 1$, then $A' = v \vee \neg v$ and so $h'(u) = \tilde{h}(A') = 1 = h(u)$;
if $h(u) = 0$, then $A' = v \wedge \neg v$ and so $h'(u) = \tilde{h}(A') = 0 = h(u)$.
Thus h and h' coincide on $\mathfrak{a}$, therefore $h = h'$, $\tilde{h}(B(A)) = \tilde{h}(A) = 0$, and $B(A) \notin T$.

For the case $v(A) = 1$, let g be any system of truth values, and define g' by

$$g'(u) = \tilde{g}(A'),$$

$$g'(w) = g(w) \quad \text{for } w \neq u.$$

If $h(u) = 1$, then $A' = u \vee \neg u$, and so $\tilde{g}(A') = 1 = g'(u) = h(u)$;
if $h(u) = 0$, then $A' = u \wedge \neg u$, and so $\tilde{g}(A') = 0 = g'(u) = h(u)$.
Thus in all cases $h(u) = g'(u)$, and so $\tilde{h}(A) = \tilde{g}'(A)$, since $\mathfrak{a}_A = \{u\}$.
Now it follows from Lemma 1 that

$$\tilde{g}'(A) = \tilde{g}((A'/u)\,A) = \tilde{g}(B(A)),$$

therefore $\tilde{g}(B(A)) = 0$, or $\tilde{g}(\neg B(A)) = 1$. Since this is true for all g, we have $\neg B(A) \in T$.

We can now prove the

2 Second completeness theorem (Syntactic)

If $A \notin T$, then $\Sigma(A)$ is inconsistent.

Proof Put $B_0 = A$.
Since $A \notin T$, we can define $B_1 = B(A)$ by Lemma 2.
$B(A) \notin T$, and so we can define $B_2 = B(B_1) = B(B(A))$.
$B_2 \notin T$, and so we can define $B_3 = B(B_2)$, etc.
Thus we construct a sequence of sentences $B_0, B_1, \ldots, B_r, \ldots$
Let $n = v(A) = v(B_0)$. If $n > 1$, then $v(B_1) = n - 1$; if $n > 2$, then $v(B_2) = n - 2$; and so on. It follows that $v(B_{n-1}) = 1$, and therefore $v(B_n) = v(B_{n+1}) = \cdots = 1$. Moreover, for every r we have $B_{r+1} = B(B_r) \in \Sigma(B_r)$; therefore $\Sigma(B_{r+1}) \subset \Sigma(B_r)$, and so $\Sigma(B_r) \subset \Sigma(A)$.
Now $\Sigma(B_n) \vdash B_n$, since $B_n \in \Sigma(B_n)$;
$\Sigma(B_n) \vdash \neg B_n$, since $B_n = B(B_{n-1})$ and $v(B_{n-1}) = 1$; thus $\neg B_n \in T$.
But this means that $\Sigma(B_n)$ is inconsistent, and therefore so is $\Sigma(A)$.

Examples $\Sigma(u) = E$.
$\Sigma(u \wedge v)$ is the set of all sentences of the form $A \wedge B$, therefore it contains the sentence $A \wedge \neg A$, which is equivalent to ω.

Remark Another interpretation of this theorem is the following: If we add an unprovable (in the primitive calculus L') sentence A to the axioms, the result is a new logical calculus, which is inconsistent.

We can make this statement more precise by introducing the concept of a demonstrative system. The definition is as follows:

Let J be a set of sentences. A *proof* from J of a sentence A is a finite sequence of sentences $A_1, \ldots, A_n = A$ such that each sentence A_i satisfies at least one of the following conditions:

(1) $A_i \in J$;
(2) there exist $j < i$, $B \in E$, and $u \in \mathfrak{a}$ such that $A_i = (B/u)\, A_j$;
(3) there exist $j < i$ and $k < i$ such that $A_k = A_j \rightarrow A_i$.

We denote the set of sentences that have a proof from J by $S(J)$.

$S(\emptyset) = \emptyset$, since a proof must contain at least one sentence of J.

$S(u) = E$ for any atomic sentence, since every sentence may be expressed as $A = (A/u)\, u$.

$S(T) = T$, for $T \subset S(T)$ by definition, and if $A \in S(T)$ one proves by induction on the length of a proof that $A \in T$.

$S(\mathscr{A}) = T$, where $\mathscr{A}$ denotes the set of the nine axioms, for in this case the concept coincides with that of a formal proof.

The following assertions are easily proved.

$J \subset S(J)$.

If $J \subset J'$, then $S(J) \subset S(J')$.

$S(S(J)) = S(J)$.

$A \in S(J)$ if and only if there exists a finite subset J^0 of J such that $A \in S(J^0)$.

If $A \to B \in S(J)$, then $B \in S(J, A)$. For $A \to B \in S(J, A)$, *a fortiori*, and $A \in S(J, A)$. The converse is false: for example, let $J = \emptyset$, $S(J) = \emptyset$; for any u and v we have $v \in S(u)$. However, it is not true that $u \to v \in S(\emptyset)$.

$S(J) = E$ if and only if $S(J) \cap \mathfrak{a} \neq \emptyset$.

We now define a *demonstrative system* to be any subset $\mathscr{D}$ of E such that $S(\mathscr{D}) = \mathscr{D} \neq E$.

For example, $\emptyset$ and T are demonstrative systems.

A demonstrative system $\mathscr{D}$ is therefore characterized by three conditions:

(a) If $A \in \mathscr{D}$ then $\Sigma(A) \subset \mathscr{D}$, i.e. $\Sigma(\mathscr{D}) = \mathscr{D}$.

(b) $\mathscr{D}$ is closed under modus ponens.

(c) $\mathscr{D} \neq E$ (which is equivalent to $\mathscr{D} \cap \mathfrak{a} = \emptyset$).

It is easy to verify that the set of all demonstrative systems, ordered by set inclusion, is inductive. Zorn's Lemma thus implies that there exist maximal demonstrative systems. We now state

3 Third completeness theorem (Syntactic)

T is a maximal demonstrative system.

Proof We already know that T is a demonstrative system. Suppose there exists another demonstrative system $\mathscr{D}$ such that $\mathscr{D} \supsetneq T$. Let $A \in \mathscr{D} - T$.

By the Second Completeness Theorem, $\Sigma(A)$ is inconsistent, i.e. $T(\Sigma(A)) = E$.

Now $A \in \mathscr{D}$; therefore $\Sigma(A) \subset \mathscr{D}$, so that $T(\Sigma(A)) \subset T(\mathscr{D})$, and so $T(\mathscr{D}) = E$.

But $T(\mathscr{D}) = \mathscr{D}$, since $\mathscr{D} \supset T$ and $\mathscr{D}$ is closed under modus ponens.

Therefore $\mathscr{D} = E$, a contradiction.

Remark The above theorem illustrates the concept of completeness with regard to the primitive axiom system $\mathscr{A}$, since for any other axiom system $\mathscr{A}'$ we have either $S(\mathscr{A}') = T$ or $S(\mathscr{A}') = E$.

Equivalence of the three completeness theorems

We have now proved three completeness theorems; the proofs show that

$$1 \Rightarrow 2 \Rightarrow 3.$$

We shall now see that these three theorems are equivalent. To this end we need a few preliminary lemmas.

3 LEMMA *If $A \in T(\Sigma(J))$, then $(B/u)\, A \in T(\Sigma(J))$ for any B and u.*

Proof Here $\Sigma(J)$ is an obvious notation for the set of multiple substitution instances of the sentences in J. We shall prove the assertion by induction on the length of a deduction of A from the hypotheses $\Sigma(J)$.

If $A \in T$, then $(B/u)\, A \in T$, and thus $(B/u)\, A \in T(\Sigma(J))$.

If $A \in \Sigma(J)$, then $(B/u)\, A \in \Sigma(J)$, and thus $(B/u)\, A \in T(\Sigma(J))$.

If A is obtained by modus ponens from C and $C \to A$, suppose the assertion true for the latter two sentences:

$$(B/u)\, C \in T(\Sigma(J)),$$

$$(B/u)\, C \to (B/u)\, A \in T(\Sigma(J)).$$

It follows that $(B/u)\, A \in T(\Sigma(J))$.

4 LEMMA $T(\Sigma(J)) = S(J, T)$.

Proof Let $A \in T(\Sigma(J))$. We reason by induction on the length of a deduction of A from the hypotheses $\Sigma(J)$:

If $A \in T$, then $A \in S(J, T)$.

If $A \in \Sigma(J)$, then $A \in S(J)$, and thus $A \in S(J, T)$.

If A is obtained by modus ponens from B and $B \to A$, where $B \in S(J, T)$ and $B \to A \in S(J, T)$, then $A \in S(J, T)$.

Now let $A \in S(T, J)$, and reason by induction on the length of a proof of A from J, T:

If $A \in J$ or $A \in T$, then $A \in \Sigma(J)$ or $A \in T$, whence $A \in T(\Sigma(J))$.

If $A = (C/u)\, B$, where $B \in T(\Sigma(J))$, we need only apply Lemma 3.

If A is obtained by modus ponens: immediate.

5 LEMMA *For any sentences A and $B \in \Sigma(A)$ and system of truth values h, there exists a system of truth values h' such that*

$$\tilde{h}'(A) = \tilde{h}(B).$$

Proof Since $B \in \Sigma(A)$, there exists a sequence of sentences $B_1 = A, \ldots, B_n = B$ such that for every i there are C_i and u_i with $B_i = (C_i/u_i)\, B_{i-1}$.

We reason by induction on n:

If $n = 1$, then $B = A$, and we take $h' = h$.

Suppose the assertion true for $n - 1$. Now $B = B_n = (C_n/u_n)\, B_{n-1}$. By Lemma 1, there exists h_1 such that $h_1(B_{n-1}) = h(B_n)$. By the induction hypothesis applied to B_{n-1} and h_1, there exists h' such that $\tilde{h}'(A) = \tilde{h}_1(B_{n-1})$. Hence $\tilde{h}'(A) = \tilde{h}(B_n)$.

Having proved these lemmas, we can now prove the required equivalences:

$3 \Rightarrow 2$: Asssume that T is a maximal demonstrative system, and let A be an unprovable sentence. By Lemma 4, $T(\Sigma(A)) = S(T, A)$. Now $A \notin T$, and so $T \subsetneqq S(T, A)$. But this implies $S(T, A) = E$, i.e. $T(\Sigma(A)) = E$, and $\Sigma(A)$ is inconsistent.

$2 \Rightarrow 1$: Let A be a universally valid sentence. Suppose $A \notin T$; then by assumption $\Sigma(A)$ is inconsistent.

Let V_h be any truth set. Since V_h is consistent, it follows that $\Sigma(A) \not\subset V_h$. Thus there exists a sentence B_h such that $B_h \in \Sigma(A)$ and $\tilde{h}(B_h) = 0$. But by Lemma 5 there exists h' such that $\tilde{h}'(A) = \tilde{h}(B_h) = 0$, which is a contradiction. Therefore $A \in T$.

Thus the three completeness theorems are indeed equivalent.

4 Note on the operators T, Σ, S

The operator S (that is, demonstrative systems) plays a less important role than the operator T (deductive systems); nevertheless, it has certain advantages. Primarily, it highlights the structure of the propositional calculus, indicating all sentences that can be constructed on the basis of given sentences utilizing only the rules of substitution and modus ponens. The construction of demonstrative systems is therefore quite general; it is of interest in other logical calculi of the same type, which use other axiom systems (while the rules of inference are the same).

On the other hand, deductive systems are bound up with a specific propositional system (their definition depends on T). At the same time, they are far more fruitful than demonstrative systems in the investigation of a given logical calculus, in particular, by dint of the fundamental theory of deduction and the simple interpretation of maximality ($A \in V$ or $\neg A \in V$).

Various relations subsist between the three operators T, Σ, S. We leave it to the reader to prove these as exercises (all the proofs use induction in the already familiar manner).

We have already seen that $T(\Sigma(J)) = S(J, T)$.

The following equalities also hold:

$$S(\Sigma(J)) = \Sigma(S(J)) = S(J),$$

$$T(S(J)) = S(T(J)).$$

Also:
$$\Sigma(T(J)) \subset T(\Sigma(J)),$$
but this may be a proper inclusion, as the following example shows.

Let $A = u \vee v \notin T$, then $T(\Sigma(A)) = E$. We claim that $\Sigma(T(A)) \neq E$. In fact:

Consider a sentence $u' \wedge v'$, where u' and v' are any two atomic sentences; $u \vee v \rightarrow u' \wedge v' \notin T$, since, for instance, $0 \vee 1 \rightarrow 0 \wedge 1 = 1 \rightarrow 0 = 0$.

Thus $u' \wedge v' \notin T(A)$ and therefore $u' \wedge v' \notin \Sigma(T(A))$, since $u' \wedge v'$ can be derived by substitutions only from another sentence $u'' \wedge v''$ of the same form.

Continuing with the above relations, note that $T(J) = T(J) \cup T$. Therefore
$$S(T(J)) = S(T(J), T) = T(\Sigma(T(J))) \subset T(T(\Sigma(J))) = T(\Sigma(J)) = S(J, T).$$
On the other hand, $J \cup T \subset T(J)$ and thus $S(J, T) \subset S(T(J))$. Finally,
$$T(S(J)) = S(T(J)) = T(\Sigma(J)) = S(J, T).$$

Another interesting question is that of *axiomatizable demonstrative systems*, that is, demonstrative systems $\mathscr{D}$ such that there exists a finite subset $\mathscr{D}^0$ with the property $\mathscr{D} = S(\mathscr{D}^0)$.

Examples T is an axiomatizable demonstrative system, since $T = S(\mathscr{A})$.

The set $\mathscr{D}$ of all sentences of the form $A \wedge B$ (i.e. sentences whose dominant logical symbol is $\wedge$) is a demonstrative system (since modus ponens cannot be applied), and it is axiomatizable, since $\mathscr{D} = S(u \wedge v)$.

There exist nonaxiomatizable demonstrative systems.

Example Let $T\neg$ denote the set of inconsistent sentences, i.e. sentences A such that $\neg A \in T$.

The set $T\neg$ is a demonstrative system:

If $A \in T\neg$, then $(B/u)\, A \in T\neg$, since $\neg(B/u)\, A = (B/u)\, \neg A \in T$.

The rule of modus ponens can never be applied—the situation $A \in T\neg$ and $A \rightarrow B \in T\neg$ cannot arise, since otherwise we would have $\neg A \in T$ and $\neg(A \rightarrow B) \in T$, therefore $\tilde{h}(A) = \tilde{h}(A) \rightarrow \tilde{h}(B) = 0$ for all h; but $0 \rightarrow 1 = 1$, and so this is impossible.

That $T\neg \neq E$ is immediate.

Now suppose the set $T\neg$ were axiomatizable, say $T\neg = S(A_1, \ldots, A_n)$, or simply $T\neg = \Sigma(A_1, \ldots, A_n)$ (since modus ponens cannot be applied).

Now $u \wedge \neg u \in T\neg$, and therefore at least one of the sentences A_i, say A_1, has the form

$$A_1 = v \wedge \neg w.$$

Similarly, $\neg\neg(u \wedge \neg u) \in T\neg$, and therefore at least one of the "axioms", say A_2, has the form $A_2 = \neg\neg(v' \wedge \neg w')$, and so on. This contradicts our assumption that the number of "axioms" is finite.

In the same way it can be verified that the set $\neg T$ of sentences of the form $\neg A$, where $A \in T$, is a nonaxiomatizable demonstrative system.

Remark $\neg T \subset T\neg$: Consider $\neg A$, where $A \in T$; then $\neg\neg A \in T$, therefore $\neg A \in T\neg$. However, the inclusion is proper, for example:

$$u \wedge \neg u \in T\neg \quad \text{but} \quad u \wedge \neg u \notin \neg T.$$

CHAPTER V

Algebraic approach: structure of Boolean rings

1 Binary operations in E/R

RECALL THAT WE have defined certain relations in the set E of sentences (for both calculi L' and L''):

A partial order relation: $A \vdash B$

or $A \to B \in T$,

and an equivalence relation: $A \vdash B$ and $B \vdash A$,

or $A \leftrightarrow B \in T$.

We shall use the notation $A \equiv B$ for this equivalence relation R, and call it the *congruence relation* (the sentences A and B are *congruent*).*

These relations may also be expressed in terms of cds, if we remember that T is the intersection of all cds and use Proposition 36 of Chapter III:

$A \vdash B$ if and only if $A \to B \in T$,

or: if and only if for every cds V, $A \to B \in V$,

or: if and only if for every cds V, if $A \in V$ then $B \in V$.

$A \equiv B$ if and only if for every cds V, if $A \in V$ then $B \in V$, and conversely.

This will enable us to prove that the congruence relation is compatible with the connectives, when the latter are regarded as operations in E.

1 LEMMA *If $A \equiv A'$ and $B \equiv B'$, then:*

$$\neg A \equiv \neg A'$$
$$A \wedge B \equiv A' \wedge B'$$
$$A \vee B \equiv A' \vee B'$$
$$A \to B \equiv A' \to B'$$
$$A \leftrightarrow B \equiv A' \leftrightarrow B'.$$

Proof Let V be a cds such that $\neg A \in V$; then $A \notin V$, therefore $A' \notin V$ and $\neg A' \in V$, and vice versa.

Let V be a cds such that $A \wedge B \in V$. Then $A \in V$ and $B \in V$, therefore $A' \in V$ and $B' \in V$, so that $A' \wedge B' \in V$; and vice versa.

* *Translator's note* The French original has *analogie*. We have not been able to find any similar use of the English word "analogy", and therefore prefer "congruence", in view of the properties proved in Lemma 1 below.

The proof for the remaining cases is analogous (use Proposition 36 of Chapter III).

For the calculus L'' we have an additional assertion:

2 Lemma *If $A \equiv A'$, then for any individuals a, b and variable x (with no occurrence in A or A'):*

$$(a/b)\,A \equiv (a/b)\,A'$$
$$\exists x\,[(x/a)\,A] \equiv \exists x\,[(x/a)\,A']$$
$$\forall x\,[(x/a)\,A] \equiv \forall x\,[(x/a)\,A'].$$

Proof If $A \leftrightarrow A' \in T$, then $(a/b)\,(A \leftrightarrow A') = (a/b)\,A \leftrightarrow (a/b)\,A' \in T$.

By one of the axioms, $\forall x\,[(x/a)\,A] \vdash (a/x)\,(x/a)\,A$,

or $\forall x\,[(x/a)\,A] \vdash A$;

but $A \vdash A'$,

therefore $\forall x\,[(x/a)\,A] \vdash A'$,

or $\forall x\,[(x/a)\,A] \vdash (a/x)\,(x/a)\,A'$.

Now the individual a occurs neither in $\forall x\,[(x/a)\,A]$ nor in $(x/a)\,A'$, so that by the principle of generalization (p. 50)

$$\forall x\,[(x/a)\,A] \vdash \forall x\,[(x/a)\,A'];$$

the converse is proved in a similar way.

The proof for $\exists x\,[(x/a)\,A]$ is analogous.

Remark 1 If $A \rangle a \langle$, then $\exists x\,[(x/a)\,A] \equiv A \equiv \forall x\,[(x/a)\,A]$; for $A = (a/x)\,(x/a)\,A \vdash A$ and a does not occur in A and $(x/a)\,A$, therefore $\exists x\,[(x/a)\,A] \vdash A$; the converse relation is trivial.

Remark 2 If $A \equiv A' \rangle a \langle$, then $\exists x\,[(x/a)\,A] \equiv A \equiv \forall x\,[(x/a)\,A]$.

Remark 3 If $A \rangle x, y \langle$, then $\exists y\,[(x/a)\,A] \equiv \exists x\,[(y/a)\,A]$.

Proof $(a/x)\,(x/a)\,A \vdash \exists x\,[(x/a)\,A]$ (axiom)

$$A \vdash \exists x\,[(x/a)\,A]$$

$$(a/y)\,(y/a)\,A \vdash \exists x\,[(x/a)\,A]$$

whence $\exists y\,[(y/a)\,A] \vdash \exists x\,[(x/a)\,A]$, since a has no occurrence in $(y/a)\,A$, $\exists x\,[(x/a)\,A]$. The proof of the converse is similar.

Remark 4 If $A \equiv A'$, let x be such that $A \rangle x \langle$ and y such that $A' \rangle y \langle$; then

$$\exists x\,[(x/a)\,A] \equiv \exists x\,[(y/a)\,A'].$$

In fact, let z be such that $A\ \rangle x, z\langle$ and $A'\ \rangle y, z\langle$; then

$$\exists x\,[(x/a)\,A] \equiv \exists z\,[(z/a)\,A] \equiv \exists z\,[(z/a)\,A'] \equiv \exists y\,[(y/a)\,A'].$$

These results make it possible to define operations in E/R. We denote the elements of E/R by $\alpha, \beta, \gamma, \ldots$ and the canonical mapping of E onto E/R by φ.

Then if $\alpha = \varphi(A)$, we can define:

$$\begin{aligned} \neg\alpha &= \varphi(\neg A)\\ \alpha \wedge \beta &= \varphi\,(A \wedge B)\\ \alpha \vee \beta &= \varphi\,(A \vee B)\\ \alpha \rightarrow \beta &= \varphi\,(A \rightarrow B)\\ \alpha \leftrightarrow \beta &= \varphi\,(A \leftrightarrow B). \end{aligned}$$

The preceding lemmas show that these operations are well defined, i.e. they are independent of the representatives A and B of the congruence classes α and β.

Henceforth we shall denote the operation $\alpha \wedge \beta$ by $\alpha\,.\,\beta$, or simply $\alpha\beta$, and call it "multiplication" instead of conjunction.

We know that T is a congruence class; we denote it by 1: $1 = \varphi(A)$, where A is any provable sentence.

Similarly, we put $0 = \neg 1$, i.e., $0 = \varphi(\neg A)$ where $A \in T$.

Remark This does not mean that $0 = \neg T$, since $\neg T$ is not a congruence class. However, we can prove that $0 = T\neg$ (the set of inconsistent sentences):

Let $A \in T\neg$ and $B \equiv A$; since $B \vdash A$ it follows that $\neg A \vdash \neg B$. But $\neg A \in T$, and so $\neg B \in T$, or $B \in T\neg$.

If $A \in T\neg$ and $B \in T\neg$, then $\neg A \in T$ and $\neg B \in T$; therefore $\neg A \vdash \neg B$ and $\neg B \vdash \neg A$. Hence $A \vdash B$ and $B \vdash A$, that is, $A \equiv B$. Thus $T\neg$ is a congruence class.

Now $\neg T \subset T\neg$, and therefore $0 = T\neg$.

For the calculus L'' we can also define:

Substitution of an individual:

$$(a/b)\,\alpha = \varphi\,((a/b)\,A), \quad \text{where} \quad \alpha = \varphi(A).$$

Quantification:

$$\exists a\alpha = \varphi\,(\exists x\,[(x/a)\,A])$$

$$\forall a\alpha = \varphi\,(\forall x\,[(x/a)\,A]),$$

where A is any representative of α and x an individual which does not occur in A.

We shall now verify the usual properties of these operations, first with regard to multiplication, disjunction, and negation.

Commutativity
$$\alpha \vee \beta = \beta \vee \alpha$$
$$\alpha\beta = \beta\alpha.$$

In fact, for any cds V:

$A \vee B \in V$ if and only if $A \in V$ or $B \in V$, which is true if and only if
$$B \vee A \in V;$$
similarly for $A \wedge B$.

Associativity
$$\alpha \vee (\beta \vee \gamma) = (\alpha \vee \beta) \vee \gamma$$
and $\alpha(\beta\gamma) = (\alpha\beta)\gamma$ by an analogous proof.

Distributivity
$$\alpha \vee (\beta\gamma) = (\alpha \vee \beta)(\alpha \vee \gamma)$$
$$\alpha(\beta \vee \gamma) = (\alpha\beta) \vee (\alpha\gamma)$$

Idempotence
$$\alpha \vee \alpha = \alpha$$
$$\alpha\alpha = \alpha \quad \text{or} \quad \alpha^2 = \alpha.$$

Negation
$$\neg(\alpha \vee \beta) = \neg\alpha \,.\, \neg\beta$$
$$\neg(\alpha\beta) = \neg\alpha \vee \neg\beta$$
$$\neg\neg\alpha = \alpha$$

Neutral elements

$\alpha \vee \neg\alpha = 1$, since $A \vee \neg A \in T$;

$\alpha \vee 1 = 1$, since if $B \in T$ then $A \vee B \in V$ for any cds V;

$\alpha \,.\, 0 = 0$

$\alpha \,.\, \neg\alpha = 0$

$\alpha \,.\, 1 = \alpha$, since if $B \in T$, $A \wedge B \in V$ if and only if $A \in V$;

$\alpha \vee 0 = \alpha.$

We define yet another operation in E/R, which we call *addition* (or *exclusive disjunction*):
$$\alpha + \beta = (\alpha \,.\, \neg\beta) \vee (\neg\alpha \,.\, \beta).$$

Note that this corresponds to a connective $+$ (hitherto unused) with the property that, for any cds V,

$A + B \in V$ if and only if either $A \in V$ and $B \notin V$
or $B \in V$ and $A \notin V$.

This operation is commutative:

$$\alpha + \beta = \beta + \alpha.$$

Note also the formula

$$\alpha + \beta = (\alpha \vee \beta)(\neg\alpha \vee \neg\beta).$$

Proof $\alpha + \beta = (\alpha . \neg\beta) \vee (\neg\alpha . \beta) = ((\alpha . \neg\beta) \vee \neg\alpha) . ((\alpha . \neg\beta) \vee \beta)$

$$= (\alpha \vee \neg\alpha) . (\neg\beta \vee \neg\alpha) . (\alpha \vee \beta) . (\neg\beta \vee \beta).$$

Since $\alpha \vee \neg\alpha = \beta \vee \neg\beta = 1$, this proves the formula.

Hence the result $\alpha + \beta = (\neg\alpha) + \beta = \alpha + (\neg\beta)$.

Addition is associative: $\alpha + (\beta + \gamma) = (\alpha + \beta) + \gamma$.

To prove this, note first that $(\alpha + \beta) + \gamma = \gamma + (\beta + \alpha)$, so that by (formally) interchanging α and γ we get $(\alpha + \beta) + \gamma$ from $\alpha + (\beta + \gamma)$. Thus, set $\delta = \beta + \gamma$. Then:

$$\alpha + (\beta + \gamma) = \alpha + \delta = (\alpha . \neg\delta) \vee (\neg\alpha . \delta)$$

$$\delta = (\beta . \neg\gamma) \vee (\neg\beta . \gamma)$$

$$\delta = (\beta \vee \gamma) . (\neg\beta \vee \neg\gamma), \quad \text{so that} \quad \neg\delta = (\neg\beta . \neg\gamma) \vee (\beta . \gamma),$$

and thus

$$\alpha + (\beta + \gamma) = (\alpha . ((\neg\beta . \neg\gamma) \vee (\beta . \gamma))) \vee (\neg\alpha . ((\beta . \neg\gamma) \vee (\neg\beta . \gamma)))$$

$$= (\alpha . \neg\beta . \neg\gamma) \vee (\alpha . \beta . \gamma) \vee (\neg\alpha . \beta . \neg\gamma) \vee (\neg\alpha . \neg\beta . \gamma),$$

and this expression is indeed invariant under interchange of α and γ.

Distributivity also holds true:

$$\alpha(\beta + \gamma) = \alpha\beta + \alpha\gamma.$$

Proof $\alpha(\beta + \gamma) = \alpha((\beta . \neg\gamma) \vee (\neg\beta . \gamma)) = (\alpha . \beta . \neg\gamma) \vee (\alpha . \neg\beta . \gamma)$

$$\begin{aligned} \alpha\beta + \alpha\gamma &= (\alpha\beta(\neg\alpha \vee \neg\gamma)) \vee ((\neg\alpha \vee \neg\beta)\alpha\gamma) \\ &= (\alpha . \beta . \neg\alpha) \vee (\alpha . \beta . \neg\gamma) \vee (\neg\alpha . \alpha . \gamma) \vee (\neg\beta . \alpha . \gamma) \\ &= (\alpha . \beta . \neg\gamma) \vee (\neg\beta . \alpha . \gamma). \end{aligned}$$

Finally, the properties pertaining to neutral elements are:

$$\alpha + 0 = \alpha, \quad \text{since} \quad \alpha + 0 = (\alpha . \neg 0) \vee (\neg\alpha . 0) = (\alpha . 1) \vee 0 = \alpha;$$

$$\alpha + \alpha = 0, \quad \text{since} \quad \alpha + \alpha = (\alpha . \neg\alpha) \vee (\neg\alpha . \alpha) = 0 \vee 0 = 0.$$

The properties of addition and multiplication proved above show that these operations define over the set E/R the structure of a commutative ring with unit element, in which each element is its own additive inverse. Moreover, the property of multiplicative idempotence shows that this is a *Boolean ring* (the general theory of this structure will be reviewed in the next section).

Remark 1 It is interesting to note that the operations $\neg \vee \rightarrow \leftrightarrow$ may all be expressed in terms of $+$ and $.$ alone:

$$\neg\alpha = \alpha + 1,$$

since $\alpha + 1 = (\neg\alpha . 1) \vee (\alpha . 0) = \neg\alpha$.

$$\alpha \vee \beta = \alpha + \beta + \alpha\beta,$$

since $\alpha \vee \beta + \alpha\beta = ((\alpha \vee \beta)(\neg\alpha \vee \neg\beta)) \vee (\neg\alpha . \neg\beta . \alpha . \beta)$ and $\neg\alpha . \neg\beta . \alpha . \beta = 0$.

$$\alpha \leftrightarrow \beta = \neg\alpha \vee \beta = \alpha + \alpha\beta + 1,$$

since $(A \rightarrow B) \leftrightarrow (\neg A \vee B) \in T$, therefore $\alpha \rightarrow \beta = \neg\alpha \vee \beta$, and

$$\begin{aligned} \neg\alpha \vee \beta = (\alpha + 1) \vee \beta &= \alpha + 1 + \beta + (\alpha + 1)\beta \\ &= \alpha + 1 + \beta + \alpha\beta + \beta \\ &= \alpha + \alpha\beta + 1. \end{aligned}$$
$$\alpha \leftrightarrow \beta = (\alpha \rightarrow \beta)(\beta \rightarrow \alpha) = \neg(\alpha + \beta) = \alpha + \beta + 1,$$

since if $A \leftrightarrow B \in V$, then $A \in V$ and $B \in V$, or $A \notin V$ and $B \notin V$; therefore $A \rightarrow B \in V$ and $B \rightarrow A \in V$, and conversely.

Thus $\alpha \leftrightarrow \beta = (\alpha \rightarrow \beta)(\beta \rightarrow \alpha) = (\neg\alpha \vee \beta)(\neg\beta \vee \alpha) = \neg(\alpha + \beta) = \alpha + \beta + 1$.

Note that $\alpha \leftrightarrow \beta = 1$ if and only if $\alpha = \beta$.

Remark 2. In fact, all the operations may be expressed in termes of one suitably chosen operator (in other words, the propositional calculus may be expressed in terms of a single connective). For example:

Peirce's arrow (1880): $\downarrow$, defined by $\alpha \downarrow \beta = \neg\alpha . \neg\beta$, for then we have:

$$\begin{aligned} \neg\alpha &= \alpha \downarrow \alpha \\ \alpha . \beta &= (\alpha \downarrow \alpha) \downarrow (\beta \downarrow \beta) \\ \alpha \vee \beta &= (\alpha \downarrow \beta) \downarrow (\alpha \downarrow \beta) \\ \alpha \rightarrow \beta &= ((\alpha \downarrow \alpha) \downarrow \beta) \downarrow ((\alpha \downarrow \alpha) \downarrow \beta) \\ \alpha \leftrightarrow \beta &= ((\alpha \downarrow \alpha) \downarrow \beta) \downarrow ((\beta \downarrow \beta) \downarrow \alpha). \end{aligned}$$

Sheffer's stroke (1921): $|$ defined by

$$\alpha \,|\, \beta = \neg\alpha \vee \neg\beta,$$

whence
$$\neg\alpha = \alpha \mid \alpha$$
$$\alpha \vee \beta = (\alpha \mid \alpha) \mid (\beta \mid \beta)$$
$$\alpha . \beta = (\alpha \mid \beta) \mid (\alpha \mid \beta)$$
$$\alpha \to \beta = \alpha \mid (\beta \mid \beta).$$

Remark 3 Since congruence is the equivalence relation associated with the partial order relation $A \vdash B$, it follows that in the quotient we obtain an *order relation* in E/R:

$$\alpha \leqslant \beta \text{ if } A \vdash B, \text{ where } \alpha = \varphi(A) \text{ and } \beta = \varphi(B).$$

This order relation may also be represented by $\alpha \to \beta = 1$, or $\alpha + \alpha\beta + 1 = 1$, i.e. $\alpha\beta = \alpha$.

Moreover, we shall see that this is a general property of Boolean rings.

Specific properties of the calculus L''

For the calculus L'', the structure of E/R is richer than that of a Boolean ring, since we have at our disposal operations of substitution and quantification. Simple relations hold between these operations; in particular:

$$\forall a\alpha = \neg\exists a(\neg\alpha), \quad \exists a\alpha = \sup_b (b/a)\,\alpha, \quad \forall a\alpha = \inf_b (b/a)\,\alpha.$$

Proof The first relation follows directly from Proposition 26 of Chapter III.

As for the second:

For any b: $\quad (b/x)\,(x/a)\,A \to \exists x\,[(x/a)\,A] \in T \quad$ (axiom)

or: $\quad (b/a)\,A \to \exists x\,[(x/a)\,A] \in T$

hence: $\quad (b/a)\,\alpha \to \exists a\alpha = 1,$

so that $(b/a)\,\alpha \leqslant \exists a\alpha$, where the order relation is that just defined.

Now let β be such that $\beta \geqslant (b/a)\,\alpha$ for all individuals b. If $\beta = \varphi(B)$, then $(b/a)\,A \to B \in T$. In other words, $(b/x)\,(x/a)\,A \vdash B$ for all b. In particular, if $b \notin I_{A,B}$, we have $\exists x\,[(x/a)\,A] \vdash B$, and it follows that $\exists a\alpha \leqslant \beta$.

Thus $\exists a\alpha$ is indeed the lowest upper bound of all the $(b/a)\,\alpha$.

The proof for $\forall a\alpha$ is similar.

Given $\alpha \in E/R$ and an individual a, we shall say that α is *independent of a* if $(b/a)\,\alpha = \alpha$ for all individuals b. Let $i(a)$ denote the set of elements α that are independent of a given individual a, and I_α the set of individuals a such that a given element α depends on a (i.e. is not independent of a).

1 Proposition *The following assertions are equivalent:*

(1) For every individual b: $(b/a)\,\alpha = \alpha$.

(2) There exists a representative A of α *such that* $a \notin I_A$.

(3) There exists an individual $b \neq a$ *such that* $(b/a)\,\alpha = \alpha$.

(4) $\exists a\alpha = \alpha$.

Proof $1 \Rightarrow 2$: If $\alpha = (b/a)\,\alpha$ where $b \neq a$, let A be any representative of α. Then $A' = (b/a)\,A$ is also a representative of α, and $a \notin I_{A'}$.

$3 \Rightarrow 2$: same proof.

$1 \Rightarrow 3$: trivial.

$2 \Rightarrow 1$: If $\alpha = \varphi(A)$ where $a \notin I_A$, then for any b we have

$$(b/a)\,\alpha = \varphi\,((b/a)\,A) = \varphi(A) = \alpha.$$

$1 \Rightarrow 4$: trivial.

$4 \Rightarrow 2$: If $\exists a\alpha = \alpha$, then $\alpha = \varphi\,(\exists x\,[(x/a)\,A])$, so that there is a representative of α which is independent of a.

Remark I_α is always a finite set; moreover, $I_\alpha = \cap I_A$, where the intersection extends over all representatives A of α.

Before going more deeply into the algebraic investigation of the calculi L' and L'', we shall consider a few results from the general theory of Boolean rings, paying particular attention to the concepts of filter and ultrafilter.

2 General theory of Boolean rings

A Boolean ring is any ring B with unit element in which multiplication is idempotent:

$$x^2 = x \quad \text{for any} \quad x \in B.$$

Thus, for any two elements x and y we have

$$\begin{aligned}(x + y)^2 &= x + y \\ &= x + xy + yx + y, \quad \text{whence} \quad xy + yx = 0.\end{aligned}$$

In particular, if we set $y = x$ we get $x + x = 0$, so that each element is its own additive inverse.

The uniqueness of the additive inverse implies that $yx = xy$, and thus any Boolean ring is commutative.

In any Boolean ring one can define additional operations, analogous to those of the logical calculi:

negation: $\neg x = x + 1;$
disjunction: $x \vee y = x + y + xy;$
conditional: $x \to y = x + xy + 1 = (\neg x) \vee y;$
biconditional: $x \leftrightarrow y = x + y + 1 = \neg(x + y) = (x \to y)(y \to x).$

It is easily verified that these operations have all the properties proved above for their analogues in E/R.

Moreover, in any Boolean ring the following equalities (corresponding to the nine axioms of the calculus L') are valid:

$$x \to (y \to x) = 1$$
$$(x \to (x \to y)) \to (x \to y) = 1$$
$$(x \to y) \to ((y \to z) \to (x \to z)) = 1$$
$$(x \leftrightarrow y) \to (x \to y) = 1$$
$$(x \leftrightarrow y) \to (y \to x) = 1$$
$$(x \to y) \to ((y \to x) \to (x \leftrightarrow y)) = 1$$
$$(\neg y \to \neg x) \to (x \to y) = 1$$
$$x \vee y \leftrightarrow (\neg x \to y) = 1$$
$$xy \leftrightarrow \neg(\neg x \vee \neg y) = 1.$$

We leave it to the reader to verify these equalities—one need only carry out the indicated computations.

A *Boolean subring* B' of a Boolean ring B is any subring of B with unit element. A necessary and sufficient condition for B' to be a Boolean subring is that B' be closed with respect to multiplication and negation, since addition may be expressed as follows:

$$x + y = (x . \neg y) \vee (\neg x . y) = \neg(\neg(x . \neg y) . \neg(\neg x . y))$$

and $1 = (\neg x) + x$.

A *homomorphism* of a Boolean ring B into a Boolean ring B' is any homomorphism preserving the unit element. A mapping f has this property if and only if

$$f(x . y) = f(x) . f(y) \quad \text{and} \quad f(\neg x) = \neg f(x),$$

for this implies that f preserves addition, and also that $f(1) = 1$.

Remark An alternative necessary and sufficient condition for f to be a homomorphism is:

$$f(x \vee y) = f(x) \vee f(y) \quad \text{and} \quad f(\neg x) = \neg f(x),$$

for $x + y = (\neg(\neg x \vee y)) \vee (\neg(x \vee \neg y))$ and $x . y = \neg(\neg x \vee \neg y)$.

It is sometimes interesting to consider a less restrictive concept, which we call a semimorphism, defined by the two conditions

$$f(0) = 0 \quad \text{and} \quad f(x \vee y) = f(x) \vee f(y).$$

Every homomorphism is clearly a semimorphism, but the converse is false (counter-example: the identically zero mapping).

Order relation The relation $xy = x$ is an order relation. Indeed:

$$xx = x;$$
$$xy = x \quad \text{and} \quad yx = y \quad \text{imply that} \quad x = y \quad \text{(commutativity)};$$
$$xy = x \quad \text{and} \quad yz = y \quad \text{imply:} \quad x = x(yz) = (xy)z = xz.$$

We denote this relation by $x \leqslant y$; every Boolean ring is ordered by this relation (which is precisely that defined above in E/R).

The order relation may also be expressed in other, equivalent forms:

$$x \vee y = y$$
$$x \to y = 1$$
$$\neg x \vee y = 1$$
$$x \,.\, \neg y = 0$$

In fact:

$x \vee y = x + y + xy$, and therefore $xy = x$ if and only if $x \vee y = y$.

$x \to y = x + xy + 1$, and therefore $xy = x$ if and only if $x \to y = 1$, or $\neg x \vee y = 1$,

Note that this relation is not a total order relation, since $x \,.\, \neg x = 0$, and thus (if $x \neq 0$ and $x \neq 1$), x and $\neg x$ are incomparable (this also shows that no Boolean ring containing more than three elements can be an integral domain).

The order relation possesses the following properties:

(1) The elements 1 and 0 are respectively the greatest and smallest elements of B, since for any x we have $x \,.\, 1 = x$ and $x \,.\, 0 = 0$.

(2) $x \vee y = \sup(x, y)$ and $x \,.\, y = \inf(x, y)$.

Proof: $x \leqslant x \vee y$, since $x \vee (x \vee y) = (x \vee x) \vee y = x \vee y$; similarly, $y \leqslant x \vee y$. Let z be such that $x \leqslant z$ and $y \leqslant z$: $x \vee z =$ and $y \vee z = z$; then

$$(x \vee y) \vee z = x \vee (y \vee z) = x \vee z = z.$$

The argument for $x.\,y$ is analogous. This implies that B is a lattice.

(3) $x \leqslant y$ if and only if $\neg y \leqslant \neg x$ (i.e. $y + 1 \leqslant x + 1$), since $\neg y \to \neg x = \neg\neg y \vee \neg x = \neg x \vee y = x \to y$.

(4) If $x \leqslant x'$ and $y \leqslant y'$, then $x \,.\, y \leqslant x' \,.\, y'$ and $x \vee y \leqslant x' \vee y'$.

This is a direct result of the properties of lowest upper bounds and greatest lower bounds.

(5) If f is a semimorphism (*a fortiori*, a homomorphism), then it is a monotonic mapping.

Proof If $x \leqslant y$, then $x \vee y = y$; therefore $f(x) \vee f(y) = f(y)$, i.e. $f(x) \leqslant f(y)$.

In particular, for any semimorphism f:

$$f(xy) \leqslant f(x) f(y),$$

since $xy \leqslant x$ and $xy \leqslant y$.

Classical examples of Boolean rings (a) The set $\mathbb{U} = \{0, 1\}$, which we have already encountered in our discussion of semantics.

(b) Let E be any set; then its power set $\mathfrak{p}(E)$ (the set of all subsets of E) is a Boolean ring with respect to the usual operations: intersection (= multiplication), union (= disjunction), complementation (= negation). The order relation is simply set inclusion.

(c) The sets E/R studied above are Boolean rings.

Filters and ultrafilters

A filter in a Boolean ring B is any subset F of B with the properties:

(1) If $x \in F$ and $y \geqslant x$, then $y \in F$.

(2) If $x \in F$ and $y \in F$, then $xy \in F$.

(3) $0 \notin F$.

Remark 1 Condition (3) simply means that $F \neq B$; equivalently, one could say that B is an improper filter.

Remark 2 Conditions (1) and (2) may be combined into a single equivalent condition: If $x \in F$ and $y \in F$, then the interval $[xy, 1]$ is contained in F.

Remark 3 The set $\{1\}$ is a filter—the smallest possible.

Remark 4 For any element $x \neq 0$, the set F of all elements y such that $y \geqslant x$ is a filter, as is easily verified. This filter is precisely the set $F = B \vee x$

(i.e. the set of all elements $y \vee x$), for

(a) if $y \geqslant x$ then $y = y \vee x$;
(b) if $y = z \vee x$, then $y \geqslant x$.

This filter is known as the *principal filter* generated by the element x (and x is its smallest element).

Remark 5 This concept may be generalized as follows.

Call a subset E of B *consistent* if no finite product of elements of E is ever zero. Given a consistent subset E, let F be the set of all elements y with the property: there exists a finite product $p = x_1 \cdots x_n$ of elements of E such that $y \geqslant p$; then F is a filter:

(1) if $y \geqslant p$ and $z \geqslant y$, then *a fortiori* $z \geqslant p$;

(2) if $y \geqslant p$ and $z \geqslant p'$, then $yz \geqslant pp'$;

(3) $0 \notin F$, since otherwise some finite product in E would be zero.

Now F is easily seen to be the smallest filter containing E. It is known as the filter *generated* by the consistent subset E.

It is obvious that any filter is consistent and is generated by itself.

The filter generated by an inconsistent subset is the improper filter.

The concept of filter is related in a simple way to that of *ideal*. In general, an ideal is a subset I of B such that

(1′) if $x \in I$ and $y \in I$, then $x + y \in I$;
(2′) if $x \in I$ and $a \in B$, then $ax \in I$.

2 PROPOSITION *In a Boolean ring, conditions (1′) and (2′) are equivalent to the following:*
(1″) if $x \in I$ and $y \leqslant x$, then $y \in I$;
(2″) if $x \in I$ and $y \in I$, then $x \vee y \in I$.

Proof If properties (1′) and (2′) hold:

Let $x \in I$ and $y \leqslant x$; then $y = xy$, and so $y \in I$. Moreover, if $x \in I$ and $y \in I$, then $x \vee y = x + y + xy$, so that $x \vee y \in I$.

Conversely, assume that properties (1″) and (2″) are valid. If $x \in I$ and $y \in I$, then we have $x \,.\, \neg y \leqslant x$, so that $x \,.\, \neg y \in I$; similarly, $\neg x \,.\, y \in I$. Therefore $x + y = (x \,.\, \neg y) \vee (\neg x \,.\, y) \in I$. Moreover, if $x \in I$ and $a \in B$, then $ax \leqslant x$ implies that $ax \in I$.

Remark 1 A third condition, (3″) $1 \notin I$, would mean simply that $I \neq B$, i.e. I is a proper ideal.

Remark 2 Conditions (1″) and (2″) may be combined into a single, equivalent condition: If $x \in I$ and $y \in I$, then the interval $[0, x \vee y]$ is contained in I.

It thus becomes clear that the concepts of filter and ideal are analogous. This analogy becomes even more obvious if we adopt the following notation: Given any subset E of B, let $\neg E$ denote the set of all elements of the form $\neg x$, where $x \in E$. Alternatively (since $x = \neg\neg x$), $\neg E$ may be defined as the set of all elements x such that $\neg x \in E$.

3 PROPOSITION *F is a filter if and only if $\neg F$ is a proper ideal.*

Proof Let F be a filter. Then if $x \in \neg F$ and $y \leqslant x$, $\neg x \leqslant \neg y$, so that $\neg y \in F$ or $y \in \neg F$. If $x \in \neg F$ and $y \in \neg F$, then $\neg x \,.\, \neg y \in F$, so that

$$\neg(\neg x \,.\, \neg y) = x \vee y \in \neg F.$$

$1 \notin F$, since $1 = \neg 0$.

The converse is analogous.

Remark If F is the principal filter $B \vee x$, then $\neg F$ is the principal ideal $B \,.\, \neg x$.

Consider the set of filters of B, ordered by set inclusion. It is immediate that this set is inductive, for if $\{F_\iota\}$ is a totally ordered family of filters, then $\cup F_\iota$ is also a filter. By Zorn's Lemma it follows that there exist maximal filters, which we shall call *ultrafilters*; to be precise: every filter is contained in an ultrafilter.

4 PROPOSITION *U is an ultrafilter if and only if $\neg U$ is a maximal ideal.*

Proof Let U be an ultrafilter. Then $\neg U$ is a proper ideal; let I be a proper ideal such that $I \supset \neg U$. Then $U \subset \neg I$. But $\neg I$ is a filter, therefore $U = \neg I$ or $I = \neg U$. Thus $\neg U$ is a maximal ideal.

Proof of the converse is analogous.

We now consider some properties of filters and ultrafilters.

5 PROPOSITION *If F is a filter, then $x \in F$ if and only if the set $F, \neg x$ is inconsistent.*

Proof If $x \in T$, then since $x \,.\, \neg x = 0$ it follows that $F, \neg x$ is inconsistent.

If $F, \neg x$ is inconsistent, then there is a finite product p of elements of F such that $p \,.\, \neg x = 0$. But this means that $x \geqslant p$, and since $p \in F$ it follows that $x \in F$.

6 Proposition *If F is a filter, then F is an ultrafilter if and only if F, x is inconsistent for any $x \notin F$.*

Proof Assume that F is an ultrafilter, and let $x \notin F$. Assume that F, x is consistent. Then it generates a filter F' such that $F' \supset F, x$, so that $F' \supsetneqq F$—a contradiction.

Now assume that F, x is inconsistent for any $x \notin F$, but that F is not an ultrafilter. Then there exists a filter $F' \supsetneqq F$. Thus there exists $x \in F'$ such that $x \notin F$, so that $F, x \subset F'$, and therefore F' is inconsistent—a contradiction.

7 Proposition *If F is a filter, then F is an ultrafilter if and only if, for any $x \in B$, either $x \in F$ or $\neg x \in F$.*

Proof By Proposition 6, F is an ultrafilter if and only if F, x is inconsistent for any $x \notin F$; and by Proposition 5, the latter statement is true if and only if, for any $x \notin F$, $\neg x \in F$.

8 Proposition *If U is an ultrafilter and $x_1 \vee x_2 \vee \cdots \vee x_n \in U$, then there exists i such that $x_i \in U$.*

Proof Suppose that $x_i \notin U$ for all i. Then $\neg x_i \in U$ for all i, so that $\neg x_1 \cdots \neg x_n \in U$, i.e. $\neg(x_1 \vee \cdots \vee x_n) \in U$. But this is a contradiction.

9 Proposition *Any filter is the intersection of all the ultrafilters that contain it.*

Proof Let $\{U_\iota\}$ be the family of ultrafilters that contain the filter F in question. Clearly, $F \subset \cap U_\iota$.

Now let $x \in \cap U_\iota$, and assume that $x \notin F$. Then $F, \neg x$ is consistent, and it therefore generates a filter F' which itself is contained in an ultrafilter U. Now clearly $U \supset F$, so that U is one of the U_ι, and we get the contradiction $\neg x \in U$.

Corollary *The intersection of the family of all ultrafilters is $\{1\}$.*

This follows from the fact that $\{1\}$ is a filter, and 1 belongs to all ultrafilters.

10 Proposition *If U is an ultrafilter, then $\neg U = \complement U$.*

Proof If $x \in \neg U$, then $\neg x \in U$, and therefore $x \notin U$; and conversely.

11 Proposition *A nonempty subset U of B is an ultrafilter if and only if it satisfies the two conditions:*

(a) $x \in U$ if and only if $\neg x \notin U$;

(b) $x \,.\, y \in U$ if and only if $x \in U$ and $y \in U$.

Proof If U is an ultrafilter, then (a) follows from Proposition 7, and (b) from the fact that $xy \leqslant x$ and $xy \leqslant y$.

Conversely, suppose that conditions (a) and (b) hold. Let $x \in U$ and $y \geqslant x$; then $xy = x \in U$, therefore $y \in U$. If $x \in U$ and $y \in U$, then $xy \in U$. Finally, since $x = x \,.\, 1$ for any $x \in U$, it follows that $1 \in U$; therefore $0 \notin U$.

Thus U is a filter, and, by (a), an ultrafilter.

Remark Property (b) may be replaced by the following condition:
(c) $x \vee y \in U$ if and only if either $x \in U$ or $y \in U$.

12 PROPOSITION *A subset U of B is an ultrafilter if and only if its characteristic function γ is a homomorphism of B into $\mathbb{U}$.*

Proof Recall that γ is defined by

$$\gamma(x) = \begin{cases} 1 & \text{if } x \in U \\ 0 & \text{if } x \notin U. \end{cases}$$

Now let U be an ultrafilter. Then $\gamma\,(xy) = 1$ if and only if $xy \in U$, i.e. if and only if $x \in U$ and $y \in U$, or $\gamma(x) = \gamma(y) = 1$. Thus $\gamma\,(xy) = \gamma(x)\,\gamma(y)$. $\gamma(\neg x) = 1$ if and only if $\neg x \in U$, i.e. if and only if $x \notin U$, or $\gamma(x) = 0$; thus $\gamma(\neg x) = \neg\gamma(x)$. It follows that γ is indeed a homomorphism.

Proof of the converse is analogous, based on the assumption that γ is a homomorphism and using conditions (a) and (b).

3 Stone's representation theorem

Any Boolean ring is isomorphic to a Boolean subring of the ring of subsets $\mathfrak{p}(X)$ of some set X.

To be precise: we shall prove that, if X is the set of all ultrafilters of the ring B, the required isomorphism is the mapping σ (which we shall call the Stone mapping) defined by

$$\sigma(x) = \text{the set of all ultrafilters containing } x.$$

Proof We first prove that σ is a homomorphism. Let $U \in \sigma(\neg x)$, $\neg x \in U$. Then $x \notin U$, so that $U \notin \sigma(x)$. It follows that $U \in \mathbf{C}\sigma(x)$. The converse follows by a similar argument, and finally $\sigma(\neg x) = \mathbf{C}\sigma(x)$.

Now let $U \in \sigma(xy)$, $xy \in U$. Then $x \in U$, and $y \in U$, so that $U \in \sigma(x)$ and $U \in \sigma(y)$. The converse is again easily proved, and it follows that $\sigma(xy) = \sigma(x) \cap \sigma(y)$.

It remains to prove that σ is one-to-one. Let $\sigma(x) = \sigma(y)$; then $\sigma(x) + \sigma(y) = \sigma(x + y) = \emptyset$, or $\sigma(\neg(x + y)) = X$. But this means that $\neg(x + y)$ belongs to all ultrafilters, and therefore, by the Corollary to Proposition 9, $\neg(x + y) = 1$, i.e. $x + y = 0$, and $x = y$.

Thus B is isomorphic to $B' = \sigma(B)$, which is a subring of $\mathfrak{p}(X)$.

Remark When we mention the Stone isomorphism, we shall always mean that defined above; however, it is not the only possible isomorphism. For example, in view of Proposition 12 we can take X to be the set of all homomorphisms of B into $\mathbb{U}$ and define $\sigma(x)$ as the set of all homomorphisms γ such that $\gamma(x) = 1$.

13 Proposition *If B is a finite Boolean ring, it is isomorphic to the entire ring* $\mathfrak{p}(X)$.

Proof We first show that in this case every filter is principal. Let $F = \{x_1, \ldots, x_n\}$ be any filter, and set $x_0 = x_1 x_2 \cdots x_n$. Then it is immediate that F is the principal filter generated by x_0.

Now consider any element of $\mathfrak{p}(X)$, say $\{U_1, \ldots, U_k\}$, and set $F = \bigcap_{i=1}^{k} U_i$. It is immediate that F is a filter; it is therefore a principal filter, generated by some element x_0.

For any i, $x_0 \in U_i$, or $U_i \in \sigma(x_0)$. Thus $\{U_1, \ldots, U_k\} \subset \sigma(x_0)$.

Conversely, let $U \in \sigma(x_0)$, $x_0 \in U$. Then $F \subset U$. Suppose that U is not one of the U_i; then U does not contain U_i for any i, and there exists $x_i \in U_i$ such that $x_i \notin U$. Now let $x = x_1 \vee x_2 \vee \cdots \vee x_k$. Then $x \geqslant x_i$ and therefore $x \in U_i$ for every i. But this implies that $x \in U$, whence it follows that one of the x_i must belong to U, which is a contradiction.

Thus $\{U_1, \ldots, U_k\} = \sigma(x_0)$, which shows that σ maps B *onto* $\mathfrak{p}(X)$.

One result of this is that the number of elements in a finite Boolean ring must be a power of 2, 2^n, where n is the number of its ultrafilters.

4 Application to logical calculi

We now return to the Boolean ring E/R. The general results just proved will enable us to interpret some of the logical concepts already introduced.

Consistency There is a very simple relation between consistency in its logical sense (Chapter III) and consistency in the algebraic sense, as defined above:

14 Proposition *Let J be a subset of E. Then J is consistent in the logical sense if and only if* $\varphi(J)$ *is a consistent subset of* E/R *in the algebraic sense.*

Proof If J is consistent, suppose that $\varphi(J)$ is inconsistent, i.e. there exists a finite product $\alpha_1 \cdots \alpha_n = 0$ in $\varphi(J)$. In other words, $\varphi(A_1 \wedge A_2 \wedge \cdots \wedge A_n) = 0$, where $\alpha_i = \varphi(A_i)$, $A_i \in J$. But $0 = \varphi(\omega)$, and so $\wedge A_i \vdash \omega$, hence $J \vdash \omega$; but this means that J is inconsistent—a contradiction.

Conversely, suppose that J is inconsistent. Then there is a finite subset J^0 of J which is inconsistent, $\wedge J^0 \vdash \omega$. If $J^0 = \{A_1, \ldots, A_n\}$, it follows that $\alpha_1 \cdots \alpha_n = 0$ where $\alpha_i = \varphi(A_i)$. Thus $\varphi(J)$ is inconsistent.

Deducibility Let J be a subset of E, and consider the set $T(J)$ of sentences deducible from the hypotheses J. Then:

15 PROPOSITION *If J is inconsistent, then $T(J) = E$, so that $\varphi(T(J)) = E/R$ (the improper filter).*

If J is consistent, then so is $\varphi(J)$, and then $\varphi(T(J))$ is the filter generated by $\varphi(J)$.

Proof Let $\alpha = \varphi(A)$; then $\alpha \in \varphi(T(J))$ if and only if $A \in T(J)$, since $T(J)$ is a union of congruence classes. Now $A \in T(J)$ if and only if there is a finite subset J^0 of J such that $\wedge J^0 \vdash A$. If $J^0 = \{A_1, \ldots, A_n\}$ and $\alpha_i = \varphi(A_i)$, the latter statement is equivalent to $\alpha_1 \cdots \alpha_n \leqslant \alpha$, where the α_i are elements of $\varphi(J)$.

Deductive systems Let Δ be a deductive system, i.e. a consistent subset such that $\Delta = T(\Delta)$. Then, by what we have just proved:

$$\varphi(\Delta) \text{ is a filter in } E/R.$$

Let $F = \varphi(\Delta)$. We claim that $\Delta = \varphi^{-1}(F)$ (though φ is not one-to-one). Indeed, if $A \in \Delta$, then $\varphi(A) \in F$, and so $A \in \varphi^{-1}(F)$. If $A \in \varphi^{-1}(F)$, then $\varphi(A) \in \varphi(\Delta)$, so that $A \in \Delta$ (which is the union of congruence classes).

Conversely, let F be a filter in E/R, and consider $\Delta = \varphi^{-1}(F)$. Then Δ is a deductive system.

Proof $\Delta \neq E$, for otherwise $\omega \in \varphi^{-1}(F)$, or $\varphi(\omega) = 0 \in F$.

$\Delta \supset T$, for if $A \in T$, then $\varphi(A) = 1 \in F$.

If $A \in \Delta$ and $A \to B \in \Delta$, then $\varphi(A) \in F$ and $\varphi(A \to B) \in F$, or $\varphi(A) + \varphi(A)\,\varphi(B) + 1 \in F$. Therefore $\varphi(A) \,.\, (\varphi(A) + \varphi(A)\,\varphi(B) + 1) \in F$. Hence $\varphi(A)\,\varphi(B) \in F$, and *a fortiori* $\varphi(B) \in F$, so that $B \in \Delta$.

It follows that Δ is indeed a deductive system.

It now follows that $F = \varphi(\Delta)$, since φ maps E into E/R. We have thus proved:

16 PROPOSITION *If Δ is a deductive system, then $F = \varphi(\Delta)$ is a filter and $\Delta = \varphi^{-1}(F)$.*

If F is a filter, then $\Delta = \varphi^{-1}(F)$ is a deductive system and $F = \varphi(\Delta)$.

Thus the correspondence $\Delta \to \varphi(\Delta)$ is a one-to-one correspondence between the set of deductive systems in E and the set of filters in E/R.

Remark The statement "Δ is a deductive system if and only if $\varphi(\Delta)$ is a filter" is not correct, for a filter may be the image of a set that is not a deductive system. Example: $\Delta = \{A\}$ where $A \in T$; this set is not a deductive system, but $\varphi(\Delta) = \{1\}$ is a filter.

On the other hand, the following statement is true: F is a filter if and only if $\varphi^{-1}(F)$ is a deductive system. The reason is that here the relation $(\varphi^{-1}(F)) = F$ is always valid.

Complete deductive systems The analogues of the statements just proved for deductive systems are particularly simple for cds:

17 PROPOSITION *If V is a cds, then $U = \varphi(V)$ is an ultrafilter. If U is an ultrafilter then $V = \varphi^{-1}U$ is a cds.*

Proof U is a filter; now let $\alpha \notin U$, $\alpha = \varphi(A)$, therefore $A \notin V$, or $\neg A \in V$. But this implies $\neg\alpha \in U$.

Conversely, if U is an ultrafilter, then $V = \varphi^{-1}(U)$ is, at any rate, a deductive system. Let $A \notin V$, then $\varphi(A) \notin U$, so that $\neg\varphi(A) = \varphi(\neg A) \in U$, and thus $\neg A \in V$.

We thus obtain a one-to-one correspondence between the set of cds in E and the set of ultrafilters in E/R.

Systems of truth-values and truth sets Let h be a system of truth values, $\tilde{h}$ its extension to E. This function $\tilde{h}$ is constant over each congruence class. For if A and B are congruent, then $A \leftrightarrow B \in T$, so that $\tilde{h}(A \leftrightarrow B) = \tilde{h}(A) \leftrightarrow \tilde{h}(B) = 1$. Thus $\tilde{h}(A) = \tilde{h}(B)$.

One can thus define the function $\bar{h}$ induced by $\tilde{h}$ on E/R, by setting $\bar{h}(\alpha) = \tilde{h}(A)$, where A is any representative of α.

The following diagram results:

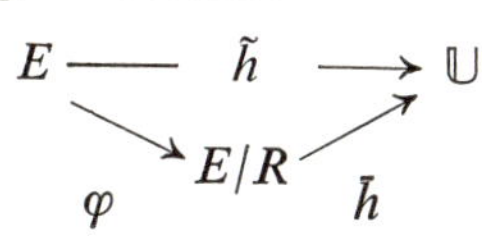

$$\tilde{h} = \bar{h}\varphi$$

The function $\bar{h}$ is a homomorphism of E/R into $\mathbb{U}$. In fact:
If $\alpha = \varphi(A)$ and $\beta = \varphi(B)$, then

$$\bar{h}(\neg\alpha) = \bar{h}(\varphi(\neg A)) = \tilde{h}(\neg A) = \neg\tilde{h}(A) = \neg\bar{h}(\alpha);$$

$$\bar{h}(\alpha\beta) = \bar{h}(\varphi(A \wedge B)) = \tilde{h}(A \wedge B) = \tilde{h}(A) \wedge \tilde{h}(B) = \bar{h}(\alpha) . \bar{h}(\alpha).$$

Thus $\bar{h}$ is the characteristic function of a certain ultrafilter U_h; moreover, h is the characteristic function of a certain truth set V_h. Then

$$U_h = \varphi(V_h) \quad \text{and} \quad V_h = \varphi^{-1}(U_h).$$

Proof If $\alpha \in U_h$, then $\bar{h}(\alpha) = 1 = \bar{h}(\varphi(A)) = \tilde{h}(A)$, so that $\alpha = \varphi(A)$, where $A \in V_h$.

If $\alpha \in \varphi(V_h)$, then $\alpha = \varphi(A)$ where $A \in V_h$; therefore $\bar{h}(\alpha) = \tilde{h}(A) = 1$, so that $\alpha \in U_h$.

Thus, if V_h is the truth set defined by the system of truth values h, then $U_h = \varphi(V_h)$ is the ultrafilter whose characteristic function is $\bar{h}$.

For the calculus L', all ultrafilters are obtained in this way, since, as we know, the concepts of cds and truth set coincide. However, in the case of the calculus L'' only certain ultrafilters are obtained in this way.

To be precise, if V_h is a truth set, then the ultrafilter $U_h = \varphi(V_h)$ satisfies the following conditions:

(i) $\exists a\alpha \in U_h$ if and only if there exists b such that $(b/a)\,\alpha \in U_h$;

(ii) $\forall a\alpha \in U_h$ if and only if $(b/a)\,\alpha \in U_h$ for every b.

Proof $\exists a\alpha = \varphi\,(\exists x\,[(x/a)\,A])$, where $\alpha = \varphi(A)$. Then $\exists a\alpha \in \varphi(V_h)$ if and only if $\exists x\,[(x/a)\,A] \in V_h$, i.e., if and only if there exists b such that $(b/x)\,(x/a)\,A \in V_h$, or $(b/a)\,A \in V_h$. Now the latter is true if and only if $(b/a)\,\alpha \in \varphi(V_h)$.

The proof of (ii) is analogous.

We claim that the conjunction of these two conditions is equivalent to the following single condition:

$\exists a\alpha \in U_h$ implies that there exists b such that $(b/a)\,\alpha \in U_h$.

Proof To prove (i) we need only note that the converse implication is trivial in any ultrafilter, since $(b/a)\,\alpha \leqslant \exists a\alpha$.

As regards (ii), recall first that $\forall a\alpha = \neg\exists a\,(\neg\alpha)$. Then:

$$\forall a\alpha \in U_h \quad \text{if and only if} \quad \exists a\,(\neg\alpha) \notin U_h,$$

i.e. if and only if $(b/a)\,(\neg\alpha) \notin U_h$ for all b, and the latter is true if and only if $(b/a)\,\alpha \in U_h$ for all b.

We shall call an ultrafilter which has the above property for any a and α a *valuating ultrafilter*.*

---

* *Translator's note* French *ultrafiltre validant*.

Thus, if V_h is a truth set (calculus L''), then $\varphi(V_h)$ is a valuating ultrafilter.

Conversely, let U be a valuating ultrafilter. We already know that $V = \varphi^{-1}(U)$ is a cds, but now we can say more:

If $\exists x\, [(x/a)\, A] \in V$, i.e. $\exists a\alpha \in U$, then there exists b such that $(b/a)\, \alpha \in U$, and it follows that $(b/a)\, A = (b/x)\,(x/a)\, A \in V$.

Thus V is a truth set (cf. Proposition 37 of Chapter III).

Certain properties relating to the calculus L''

18 Proposition *Given two individuals a and b, the mapping $\alpha \to (b/a)\, \alpha$ is a homomorphism of Boolean rings.*

Proof Let $\alpha = \varphi(A)$ and $\beta = \varphi(B)$.

$(b/a)\,(\neg\alpha) = \varphi((b/a) \neg A) = \varphi(\neg(b/a)A) = \neg(b/a)\,\alpha;$

$(b/a)\,(\alpha\beta) = \varphi((b/a)\,(A \wedge B)) = \varphi((b/a)A \wedge (b/a)B) = (b/a)\,\alpha\,.\,(b/a)\,\beta.$

In particular, it follows that (b/a) is a monotonic mapping.

19 Proposition *Given an individual a, the set $i(a)$ of all elements α independent of a is a Boolean subring of E/R.*

Proof If $\alpha \in i(a)$, then for any b, $\alpha = (b/a)\,\alpha$, whence it follows that $\neg\alpha = (b/a)\,(\neg\alpha)$, and so $\neg\alpha \in i(a)$.

If $\alpha \in i(a)$ and $\beta \in i(a)$, then for any b, $\beta = (b/a)\,\beta$, so that $\alpha\beta = (b/a)\,(\alpha\beta)$ and $\alpha\beta \in i(a)$.

20 Proposition *Moreover, for any b, $i(a) = (b/a)\,(E/R) = \exists a\,(E/R)$.*

This is an immediate result of Proposition 1 of this chapter.

21 Proposition *Every mapping $\alpha \hookrightarrow \exists a\alpha$ is a semimorphism.*

Proof $\exists a0 = 0$, since $(b/a)\, 0 = 0$.

We now show that $\exists a$ is a monotonic mapping: If $\alpha \leqslant \beta$, then $(b/a)\,\alpha \leqslant (b/a)\,\beta \leqslant \exists a\beta$ for any b, and so $\exists a\alpha \leqslant \exists a\beta$.

Finally, we prove the property $\exists a\,(\alpha \vee \beta) = \exists a\alpha \vee \exists a\beta$:

$$\alpha \leqslant \alpha \vee \beta \quad \text{and} \quad \beta \leqslant \alpha \vee \beta.$$

Therefore $\exists a\alpha \leqslant \exists a\,(\alpha \vee \beta)$ and $\exists a\beta \leqslant \exists a\,(\alpha \vee \beta)$, so that $\exists a\alpha \vee \exists a\beta \leqslant \exists a\,(\alpha \vee \beta)$.

Now $\alpha \leqslant \exists a\alpha$ and $\beta \leqslant \exists a\beta$, and therefore $\alpha \vee \beta \leqslant \exists a\alpha \vee \exists a\beta$, and for any b:

$$(b/a)\,(\alpha \vee \beta) \leqslant (b/a)\,(\exists a\alpha \vee \exists a\beta) = \exists a\alpha \vee \exists a\beta.$$

Therefore $\exists a\,(\alpha \vee \beta) \leqslant \exists a\alpha \vee \exists a\beta$.

Similarly $\exists a\,(\exists a\alpha) = \exists a\alpha$ and $\exists a\,(\neg\exists a\alpha) = (\neg\exists a\alpha$, since $\exists a\alpha \in i(a)$.

22 PROPOSITION $\exists a\,(\alpha \,.\, \exists a\beta) = \exists a\alpha \,.\, \exists a\beta$.

Proof $\alpha \leqslant \exists a\alpha$, so that $\alpha \,.\, \exists a\beta \leqslant \exists a\alpha \,.\, \exists a\beta$.

Hence $\exists a\,(\alpha \,.\, \exists a\beta) \leqslant \exists a\alpha \,.\, \exists a\beta$.

Conversely: $\alpha \leqslant \alpha \vee \neg\exists a\beta = (\alpha \,.\, \exists a\beta) \vee \neg\exists a\beta$,

whence $\exists a\alpha \leqslant \exists a\,(\alpha \,.\, \exists a\beta) \vee \neg\exists a\beta$,

or $\neg\exists a\alpha \vee \neg\exists a\beta \vee \exists a\,(\alpha \,.\, \neg\exists a\beta) = 1$,

and this implies that $\exists a\alpha \,.\, \exists a\beta \leqslant \exists a\,(\alpha \,.\, \exists a\beta)$.

These propositions show that each operator $\exists a$ is a *monadic quantifier* (cf. Halmos, *Algebraic Logic*).

CHAPTER VI

Topological approach: structure of Boolean spaces

1 General theory of Boolean spaces

In this section we present certain material from topology, which we shall use to complete our study of Boolean rings and then apply to the logical calculi.

We shall call a subset of a topological space *clopen* if it is both closed and open. Recall that a topological space X is said to be *compact* if it is a Hausdorff space and every covering of X by open sets contains a finite subcovering.

Lemma *If X is a compact topological space, the following four assertions are equivalent:*

(1) For any point x, the intersection of all clopen sets containing x is $\{x\}$.

(2) For any pair of distinct points x, y there exists at least one clopen set V such that $x \in V$ and $y \notin V$.

(3) X is generated by the family of its clopen sets, i.e., every open set is the union of clopen sets (it is evident that the union of clopen sets is an open set).

(4) X is totally disconnected, i.e., the connected component of any point x is the singleton $\{x\}$.

Proof $1 \Rightarrow 2$: trivial.

$2 \Rightarrow 3$: Let Ω be any open set; then its complement $F = \complement\Omega$ is closed. Let x be any point of Ω; then for any $y \in f$, $y \neq x$, so that there is a clopen set V_y such that $x \notin V_y$ and $y \in V_y$.

Thus $F \subset \bigcup V_y$ [where y runs through all elements of F]. But since F is closed, it is compact, and we can thus find a finite covering:

$$F \subset W, \quad \text{where} \quad W = \Delta V_{y1} \cup \cdots \cup V_{y_n}.$$

W is also a clopen set, and $x \notin W$.

It follows that Ω is the union of all its clopen subsets. Indeed, Ω contains this union trivially. Conversely, if $x \in \Omega$, let W be the clopen set constructed above; then $\complement W$ is also clopen, it contains x, and $\complement W \subset \complement F = \Omega$.

$3 \Rightarrow 4$: Let $x \in X$ and let $C(x)$ be its connected component (which is a closed set). Suppose there exists a point $y \neq x$ such that $y \in C(x)$.

Since X is Hausdorff, there exist an (open) neighborhood $\mathscr{V}_1$ of x and a neighborhood $\mathscr{V}_2$ of y such that $\mathscr{V}_1 \bigcap \mathscr{V}_2 = \emptyset$. But since any open set is a union of clopen sets, it follows ***a fortiori*** that there exists a clopen set V such that $x \in V$ and $y \notin V$.

Now consider the sets $A = V \cap C(x)$ and $B = \complement V \cap C(x)$. Both these sets are closed in $C(x)$. $A \neq \emptyset$, since $x \in A$, and $B \neq \emptyset$, since $y \in B$; $A \cap B = \emptyset$, and $A \cup B = C(x)$. This contradicts the assumption that $C(x)$ is connected. This contradiction shows that $C(x) = \{x\}$.

$4 \Rightarrow 1$: Let $\{V_\iota\}$ be the family of all clopen sets containing a given point x, and set $F = \bigcap V_\iota$. F is a closed set, and $x \in F$.

Suppose F is disconnected, i.e., there exist two sets G and G', open in F, which form a partition of F; one of these sets, say G, contains x.

Since the space X is compact, there are two open subsets H and H' such that

$$G \subset H, \quad G' \subset H', \quad H \cap H' = \emptyset.$$

$F = G \cup G' \subset H \cup H'$; therefore $\complement(H \cup H') \subset \complement F = \bigcup \complement V_\iota$. But $\complement(H \cup H')$ is compact (since it is closed), and we can thus find a finite covering:

$$\complement(H \cup H') \subset \complement V_1 \cup \cdots \cup \complement V_n, \quad \text{where} \quad H \cup H' \supset V_1 \cup \cdots \cup V_n.$$

The intersection $V_1 \cap \cdots \cap V_n$ is a certain clopen set, say V_0, and it contains x.

Now consider the sets $W = H \cap V_0$ and $W' = H' \cap V_0$. Both of these sets are open (in V_0), since H and H' are open. Moreover, W and W' are complements of each other relative to V_0, since $H \cup H' \supset V_0$; thus W is also closed in V_0, and so it is closed. It follows that W is clopen; but $x \in W$, therefore $F \supset W$.

Now $G' \neq \emptyset$, so that there is some $y \in G'$. But then $y \in F$, $y \in H'$, and $y \in V_0$; therefore $y \in W$ and $y \in W'$, which is a contradiction.

Thus F is connected, and it follows that $F = \{x\}$.

DEFINITION *A topological space X is called a Boolean space if it satisfies any (therefore all) of the above conditions.*

In practice, the most frequently used condition is (3), though conditions (1) and (2) are also very useful.

Remark Given any topological space X, the set of all its clopen sets is a *Boolean subring* B of $\mathfrak{p}(X)$, since the complement of a clopen set is also clopen and the intersection of two clopen sets is clopen.

The set $\{V_\iota\}$ of all clopen sets containing a point x is then an *ultrafilter* in the Boolean ring B:

(a) If W is a clopen set such that $W \supset V_\iota$, then $x \in W$;

(b) the intersection of two clopen sets containing x is a clopen set containing x;

(c) $\varnothing$ does not contain x;

(d) if W is a clopen set that does not contain x, then $\complement W$ is a clopen set containing x.

1 PROPOSITION *In a Boolean space every ultrafilter of the ring B is obtained in this way.*

Proof Let $\{V_\iota\}$ be any ultrafilter, and set $F = \bigcap V_\iota$.

$F \neq \varnothing$, since otherwise there would be a finite subfamily of $\{V_\iota\}$ such that $V_1 \cap \cdots \cap V_n = \varnothing$; but $V_1 \cap \cdots \cap V_n \in \{V_\iota\}$—a contradiction.

Now let $x \in F$. Any clopen set in the family $\{V_\iota\}$ contains x. Conversely, if W is a clopen set containing x then $W \in \{V_\iota\}$, since otherwise $\complement W \in \{V_\iota\}$ and $x \notin \complement W$.

Finally, suppose that F contains two distinct points x and y. Then, by condition (2), there exists a clopen set W such that $x \in W$ and $y \notin W$, so that $y \in \complement W$. But then W and $\complement W$ both along to the family $\{V_\iota\}$, which is a contradiction.

Thus F reduces to a singleton $\{x\}$, and it follows that $\{V_\iota\}$ is the set of all clopen sets containing x.

Examples of Boolean spaces Any discrete finite space is a Boolean space (any subset is clopen). In particular, we shall encounter the discrete two-element space $\{0, 1\}$, which we denote, as before, by $\mathbb{U}$.

Recall the definition of *topological product:*

Let $\{X_\iota\}$ be a family of topological spaces. In the cartesian product $X = \prod X_\iota$ we define the concept of *elementary set:* any subset of the form $\prod \Omega_\iota$, where each Ω_ι is open (in X_ι) and $\Omega_\iota = X_\iota$ for all but a finite number of indices.

The product topology in X is the topology whose base is the family of elementary sets, i.e., the open sets are precisely the unions of elementary sets.

We also recall two well-known theorems: The product of compact spaces is a compact space, and the connected component of a point $x = \{x_\iota\}$ of X is the product of the connected components of the point x_ι in X_ι.

It follows immediately that any product of Boolean spaces is a Boolean space.

Special case In the sequel we shall have occasion to deal with the case where all the factor spaces are $\mathbb{U}$. The product is then $\mathbb{U}^I$, where I is some set. Such a space is known as a *Cantor space*, and it follows from the above arguments that it is a Boolean space.

Note that every elementary set in a Cantor space is clopen: $V = \prod \Omega_\iota$ is open by definition, and it is closed since each Ω_ι is closed (discrete topology in $\mathbb{U}$).

The converse is false: a clopen set need not be an elementary set.

Boolean ring and Boolean space as dual concepts

Let B be any Boolean ring (denote its elements by $\alpha, \beta, \gamma, \ldots$), and X the set of all its ultrafilters (which we denote by $x, y, z, \ldots$).

We are already acquainted with the Stone isomorphism σ of B into $\mathfrak{p}(X)$, defined by

$$\sigma(\alpha) = \text{the set of ultrafilters } x \text{ such that } \alpha \in x,$$

and B is isomorphic to $B' = \sigma(B)$, which is a subring of $\mathfrak{p}(X)$.

Let us use this to construct a topology in X. Let $\mathcal{O}$ be the family of subsets of X which are unions of elements of B'. Then $\mathcal{O}$ satisfies the usual axioms of a topology in X:

(a) Any union of sets of $\mathcal{O}$ is again in $\mathcal{O}$.

(b) If Ω_1 and Ω_2 are in $\mathcal{O}$, then so is their intersection; for if

$$\Omega_1 = \bigcup_\iota \sigma(\alpha_\iota) \quad \text{and} \quad \Omega_2 = \bigcup_k \sigma(\beta_k),$$

then $\Omega_1 \cap \Omega_2 = \bigcup_{\iota,k} (\sigma(\alpha_\iota) \cap \sigma(\alpha_k)) = \bigcup_{\iota,k} \sigma\,(\alpha_\iota \,.\, \alpha_k)$.

(c) $X \in \mathcal{O}$, since $X = \sigma(1)$.

We may thus regard $\mathcal{O}$ as a family of open sets defining a topology in X.

2 Proposition *With the topology $\mathcal{O}$, X is a Boolean space.*

Proof We must first prove that X is a Hausdorff space. Let x and y be two distinct ultrafilters. Then there exists α such that $\alpha \in x$ and $\alpha \notin y$, i.e. $\neg\alpha \in y$. Now $x \in \sigma(\alpha)$ and $\sigma(\alpha)$ is open, and therefore $\sigma(\alpha)$ is a neighborhood of x. Similarly, $y \in \sigma(\neg\alpha)$ and $\sigma(\neg\alpha)$ is open, so that $\sigma(\neg\alpha)$ is a neighborhood of y. Moreover: $\sigma(\alpha) \cap \sigma(\neg\alpha) = \sigma\,(\alpha \,.\, \neg\alpha) = \sigma(0) = \emptyset$.

We continue the proof by showing that the family of clopen sets in X is a base for the topology. In fact, each element $\sigma(\alpha)$ of B' is open by definition, but it is also closed, for $\complement\, \sigma(\alpha) = \sigma(\neg\alpha) \in B'$. Thus each element of B' is clopen and every open set is the union of clopen sets.

Remark We shall see below that, conversely, every clopen set is an element of B' (this is not necessary for the present proof).

To complete the proof we must show that X is compact. Since B' is a base for the topology of X, it will suffice to consider a covering of X by elements of B', say:

$$X = \bigcup \sigma(\alpha_l).$$

First, we claim that the set of all elements $\neg\alpha_l$ is not consistent; otherwise, it would generate a filter, and would thus be contained in some ultrafilter x. But $x \in X$, and there therefore exists l_0 such that $x \in \sigma(\alpha_{l_0})$, so that $\alpha_{l_0} \in x$ and $\neg\alpha_{l_0} \in x$, which is a contradiction.

Since the family $\{\neg\alpha_l\}$ is inconsistent, there is a finite product of its elements which vanishes: $\neg\alpha_1 \cdots \neg\alpha_n = 0$, or $\alpha_1 \vee \cdots \vee \alpha_n = 1$. Thus

$$X = \sigma(\alpha_1) \cup \cdots \cup \sigma(\alpha_n),$$

and this is a finite subcovering of X, as required.

This completes the proof that X is a Boolean space.

We now show (as mentioned above) that B' is precisely the set of clopen sets of X.

Let V be a clopen set in X. Since V is open, it has the form $V = \bigcup \sigma(\alpha_l)$; since it is closed, we have $\complement V = \bigcup \sigma(\beta_k)$.

Now X is covered by the set consisting of all $\sigma(\alpha_l)$ together with all $\sigma(\beta_k)$; we can therefore find a finite subcovering of X:

If this subcovering contains no $\sigma(\alpha_l)$, then

$$V = X = \sigma(1) \in B'.$$

If it contains no $\sigma(\beta_k)$, then

$$V = \emptyset = \sigma(0) \in B'.$$

Finally, if it contains $\sigma(\alpha_1), \ldots, \sigma(\alpha_n), \sigma(\beta_1), \ldots, \sigma(\beta_m)$, then

$$V = \sigma(\alpha_1) \cup \cdots \cup \sigma(\alpha_n) = \sigma(\alpha_1 \vee \cdots \vee \alpha_n) \in B'.$$

Thus, with each Boolean ring B we have associated a certain Boolean space X (the set of ultrafilters of B) whose family of clopen sets coincides with $\sigma(B)$. X is known as the *Boolean space dual to B*, and we write $X = B^0$.

This yields a topological interpretation of Stone's Theorem: Every Boolean ring is isomorphic to the ring of clopen sets of its dual space.

Conversely, with each Boolean space X we can associate the Boolean ring of its clopen sets; this ring is called the ***Boolean ring dual to*** X, and we write $B = {}^0X$.

Remarks Let B be a Boolean ring, $X = B^0$ its dual space, and $B' = {}^0X = {}^0B^0$ the dual ring of X. B and B' are different, though they are isomorphic, since B' is $\sigma(B)$.

Similarly, let X be a Boolean space, $B = {}^0X$ its dual ring, and $X' = B^0 = {}^0X^0$ the dual space of B. X and X' are different, but they are homeomorphic, as we shall now prove.

Let θ be the mapping of X into X' defined as follows:

$\theta(x) =$ the set x' of clopen sets of X containing x (by the remark on p. 93 we know that x' is indeed an ultrafilter in B).

θ maps X onto X', since every ultrafilter in B is the set of clopen sets containing some point (by Proposition 1).

θ is one-to-one, since if $x \neq y$ there is a clopen set V containing x but not y.

It remains to prove that both θ and θ^{-1} are continuous. Let V' be a clopen set in X', i.e., $V' = \sigma(V)$, where V is a clopen set in X and σ is the Stone mapping of B. Then

$$\begin{aligned}\theta^{-1}(V') &= \{x \mid \theta(x) \in \sigma(V)\}\\ &= \{x \mid V \in \theta(x)\}\\ &= \{x \mid x \in V\} = V,\end{aligned}$$

and thus, finally,

$$\theta^{-1}(\theta(V)) = V \quad \text{and} \quad \theta(V) = \sigma(V),$$

which proves the required continuity.

We can thus replace any algebraic discussion of a Boolean ring by a topological discussion of its dual space, and vice versa. This concept of duality may also be extended to such concepts as homomorphisms, semimorphisms, etc. (see, e.g., P. Halmos, *Lectures on Boolean Algebra* and *Algebraic Logic*).

Another possible interpretation of the topological viewpoint makes use of continuous functions.

Let B be a Boolean ring and X its dual space. The elements $\alpha \in B$ are in one-to-one correspondence with the clopen sets $\sigma(\alpha)$ in X. Now the clopen set $\sigma(\alpha)$, as a subset of X, is uniquely determined by its characteristic function f_α, which is in effect a mapping of the Boolean space X into the discrete Boolean space $\mathbb{U}$. It is easy to see that this mapping is continuous.

In fact, let Ω be an open set (i.e., any subset) of $\mathbb{U}$; then:

If $\Omega = \emptyset, \mathbb{U}, \{1\}, \{0\}$, then $f_\alpha^{-1}(\Omega) = \emptyset, X, \sigma(\alpha), \sigma(\neg\alpha)$, respectively, and all the latter are clopen sets.

Conversely, let f be a continuous mapping of X into $\mathbb{U}$. Then, since $\{1\}$ is a clopen set in $\mathbb{U}$, it follows that $f^{-1}(1)$ is a clopen set in X, so that there exists α such that $f^{-1}(1) = \sigma(\alpha)$ and thus $f = f_\alpha$.

Note that the set of continuous mappings of X into $\mathbb{U}$ is a subset of the set $\mathscr{F}(X, \mathbb{U})$ of *all* mappings of X into $\mathbb{U}$. Now this latter set can be made into a Boolean ring, defining $\neg f$ by $(\neg f)(x) = \neg f(x)$ and fg by $(fg)(x) = f(x)\,g(x)$.

The preceding argument then shows that the set of all continuous functions is a Boolean subring of $\mathscr{F}(X, \mathbb{U})$, which is isomorphic to B.

Remark The dual space X of a Boolean ring B is the set of ultrafilters of B. Now we know that it may also be regarded as the set of homomorphisms of B into $\mathbb{U}$, and X is thus a subset of $\mathscr{F}(B, \mathbb{U})$, which is simply the product space $\mathbb{U}^B$. The reader may prove, as an exercise, that the topology defined above in X is precisely the topology induced in X by that of the Cantor space $\mathbb{U}^B$. It follows that every Boolean space is homeomorphic to a subspace of a Cantor space.

2 Applications to propositional calculus

This section concerns the calculus L' alone.

Consider the Boolean ring E/R defined previously, and let X be its dual space, i.e., the set of all ultrafilters in E/R.

If x is an ultrafilter, we know (p. 88) that its characteristic function (a homomorphism of E/R into $\mathbb{U}$) has the form $\bar{h}$, i.e., it is induced on E/R by some system of truth values h. Consider the correspondence Γ defined by

$$\Gamma(x) = h, \text{ where } \bar{h} \text{ is the characteristic function of } x.$$

The Γ is a one-to-one mapping of X onto $\vec{\mathfrak{a}}$. For we already know that it maps X onto $\vec{\mathfrak{a}}$. It is one-to-one, since if x and x' are two different ultrafilters and $\bar{h}$ and $\bar{h}'$ their characteristic functions, we must have $h \neq h'$, since $h = h'$ would imply $\bar{h} = \bar{h}'$, whence $\bar{h} = \bar{h}'$.

Remark It follows that X is equipollent to $\vec{\mathfrak{a}}$.

In particular, were $\vec{\mathfrak{a}}$ a finite set containing n elements (which is not the case), there would be 2^n ultrafilters in E/R, and thus E/R would contain 2^{2^n} elements.

The set $\vec{\mathfrak{a}}$ may be identified with the product set $\mathbb{U}^{\mathfrak{a}}$, by identifying each function h with the family $\{h(u)\}_{u \in \mathfrak{a}}$ of its values.

3 Proposition *The dual space X of E/R is homeomorphic to the Cantor space $\mathbb{U}^{\mathfrak{a}}$.*

Proof It will suffice to prove that the mapping Γ defined above is continuous. Consider a nonempty elementary set of $\mathbb{U}^{\mathfrak{a}}$:

$$V = \prod_{u \in \mathfrak{a}} \Omega_u$$

$\Omega_u = \mathbb{U}$ except for a finite number of indices.

If $\Omega_u = \mathbb{U}$ for all u, then $V = \mathbb{U}^{\mathfrak{a}}$ and $\Gamma^{-1}(V) = X$.

If $\Omega_u \neq \mathbb{U}$ for indices $u_1, \ldots, u_n$, let us associate a sentence with each index u_i, as follows:

$$\begin{aligned} A_i &= u_i \quad \text{if} \quad \Omega_{u_i} = \{1\}; \\ &= \neg u_i \quad \text{if} \quad \Omega_{u_i} = \{0\}. \end{aligned}$$

Consider the sentence $A = A_1 \wedge \cdots \wedge A_n$ (with any distribution of parentheses). If $x \in \Gamma^{-1}(V)$, then $\Gamma(x) = h \in V$, and so

$$h(u_i) = \begin{cases} 1 & \text{if} \quad A_i = u_i \\ 0 & \text{if} \quad A_i = \neg u_i, \end{cases}$$

so that in all cases $\tilde{h}(A_i) = 1$ for all i, and therefore $\tilde{h}(A) = 1$.

Conversely, if $\tilde{h}(A) = 1$ and x is the ultrafilter defined by h, then $x \in \Gamma^{-1}(V)$.

Setting $\alpha = \varphi(A)$, we have $\tilde{h}(A) = \bar{h}(\alpha)$, and so $\tilde{h}(A) = 1$ if and only if $\alpha \in x$, i.e. $x \in \sigma(\alpha)$.

It follows that

$$\Gamma^{-1}(V) = \sigma(\alpha).$$

Since the inverse image of every elementary set is a clopen set, it follows that the mapping Γ is indeed continuous, and it is therefore a homeomorphism (the continuity of the inverse follows from the fact that the spaces are compact).

Another interesting mapping is the mapping Λ of E/R into $\mathfrak{p}(\vec{\mathfrak{a}})$ defined by

$$\Lambda(\alpha) = \text{the set of all } h \text{ such that } \bar{h}(\alpha) = 1.$$

Now $\bar{h}(\alpha) = 1$ if and only if α belongs to the ultrafilter x such that $h = \Gamma(x)$, i.e. if and only if $x \in \sigma(\alpha)$ where $h = \Gamma(x)$, which in turn is equivalent to $h \in \Gamma(\sigma(\alpha))$. Thus:

$$\Lambda(\alpha) = \Gamma(\sigma(\alpha)).$$

The function Λ is a monomorphism of E/R into the Boolean ring $\mathfrak{p}(\vec{\mathfrak{a}})$:

$$\Lambda(\neg\alpha) = \Gamma(\sigma(\neg\alpha)) = \Gamma(\complement\sigma(\alpha)) = \complement\,\Gamma(\sigma(\alpha)) = \complement\,\Lambda(\alpha);$$

$$\Lambda(\alpha\beta) = \Gamma(\sigma(\alpha\beta)) = \Gamma(\sigma(\alpha) \cap \sigma(\beta)) = \Lambda(\alpha) \cap \Lambda(\beta).$$

Λ is one-to-one, for if $\alpha \neq \beta$, then $\sigma(\alpha) \neq \sigma(\beta)$, so that $\Lambda(\alpha) \neq \Lambda(\beta)$.

Note, however, that Λ does not map E/R *onto* $\mathfrak{p}(\vec{\mathfrak{a}})$. For, let $\alpha = \varphi(A)$; we know that the value of $\tilde{h}(A) = \bar{h}(\alpha)$ depends only on the values of h on $\mathfrak{a}_A$. But $\mathfrak{a}$ is infinite, and thus if $\alpha \neq 0$ and $\alpha \neq 1$, $\Lambda(\alpha)$ must be an infinite set with an infinite complement.

To derive a more precise topological result, we introduce the following definition:

Lef $\mathscr{B}$ be a subset of $\vec{\mathfrak{a}}$. We shall say that a sentence A is $\mathscr{B}$-true [$\mathscr{B}$-false] if it is h-true [h-false] for any $h \in \mathscr{B}$.

LEMMA *If $\mathscr{B}$ and $\mathscr{C}$ are two finite disjoint subsets of $\vec{\mathfrak{a}}$, then there is at least one sentence A which is $\mathscr{B}$-true and $\mathscr{C}$-false.*

Proof If $\mathscr{B} = \mathscr{C} = \varnothing$, we take A to be any sentence.

If $\mathscr{B} = \varnothing \neq \mathscr{C}$, $A = u \wedge \neg u$ (where u is any atomic sentence).

If $\mathscr{B} \neq \varnothing = \mathscr{C}$, $A = u \vee \neg u$.

If $\mathscr{B} \neq \varnothing$ and $\mathscr{C} \neq \varnothing$, let $\mathscr{B} = \{h_1, \ldots, h_n\}$, $\mathscr{C} = \{g_1, \ldots, g_p\}$. If $n = p = 1$: $\mathscr{B}$ and $\mathscr{C}$ are disjoint, so $h_1 \neq g_1$. Therefore, there exists $u \in \mathfrak{a}$ such that $h_1(u) \neq g_1(u)$. If $h_1(u) = 1$ and $g_1(u) = 0$, we take $A = u$, and otherwise $A = \neg u$.

If $n \geqslant 1$ and $p = 1$, then, by the preceding case, there exists a sentence A_1 which is h_1-true and g_1-false, ..., and there exists a sentence A_n which is h_n-true and g_1-false. We then take $A = A_1 \vee \cdots \vee A_n$.

If $n \geqslant 1$ and $p \geqslant 1$, then, by the preceding case, there exists a sentence A_1 which is $\mathscr{B}$-true and g_1-false, ..., there exists a sentence A_p which is $\mathscr{B}$-true and g_p-false. We then take $A = A_1 \wedge \cdots \wedge A_p$.

If we now identify the set $\mathfrak{p}(\vec{\mathfrak{a}})$ with the product set $\mathbb{U}^{\vec{\mathfrak{a}}}$, we have:

4 PROPOSITION *$\Lambda(E/R)$ is everywhere dense in the Cantor space $\mathbb{U}^{\vec{\mathfrak{a}}}$.*

Proof Consider a nonempty elementary set of $\mathbb{U}^{\vec{\mathfrak{a}}}$: $V = \prod_{h \in \vec{\mathfrak{a}}} \Omega_h$.

Let $\mathscr{B}$ be the set of h such that $\Omega_h = \{1\}$, and $\mathscr{C}$ the set of h such that $\Omega_h = \{0\}$. These two sets are finite and disjoint; therefore, by the lemma, there exists a sentence which is $\mathscr{B}$-true and $\mathscr{C}$-false, say A. Let $\alpha = \varphi(A)$.

As an element of $\mathbb{U}^{\vec{\mathfrak{a}}}$, $\Lambda(\alpha)$ is identified with $\{\lambda_h\}_{h \in \vec{\mathfrak{a}}}$, where

$$\lambda_h = \begin{cases} 1 & \text{if } h \in \Lambda(\alpha) \\ 0 & \text{otherwise.} \end{cases}$$

If $h \in \mathscr{B}$, then since A is h-true it follows that $h \in \Lambda(\alpha)$, therefore $\lambda_h = 1$ and $\lambda_h \in \Omega_h$.

If $h \in \mathscr{C}$, then since A is h-false, it follows that $h \notin \Lambda(\alpha)$, therefore $\lambda_h = 0$ and $\lambda_h \in \Omega_h$.

Thus $\Lambda(\alpha) \in V$.

But this means that every elementary set contains at least one element of $\Lambda(E/R)$, and this proves the assertion.

Remark $\Lambda(E/R)$ is the set of clopen sets of $\mathbb{U}^{\mathfrak{a}}$.

Proof $\Lambda(\alpha) = \Gamma(\sigma(\alpha))$ is a clopen set, since Γ is a homeomorphism.

Conversely, if V is a clopen set of $\mathbb{U}^{\mathfrak{a}}$, then $\Gamma^{-1}(V)$ is a clopen set of X, so that $\Gamma^{-1}(V) = \sigma(\alpha)$, whence $V = \Gamma(\sigma(\alpha)) = \Lambda(\alpha)$.

CHAPTER VII

Problems of completeness in the predicate calculus

1 More about truth sets

This chapter concerns only the predicate calculus.

Recall that a *truth set* V for the calculus L'' is a cds satisfying the additional condition:

$\exists x\,[f^x]$ implies that there exists an individual a such that $(a/x)f^x \in V$; or:

$$(a/x)f^x \in V \text{ for all individuals } a \text{ implies } \forall x\,[f^x] \in V.$$

(By Proposition 26 of Chapter III we know that these two properties are equivalent.)

We shall now define some other, broader concepts.

(1) Given a sentence A and an individual a, we shall call a cds V $(\exists, A, a)$-*valuating** if either $\exists x\,[(x/a)\,A] \notin V$ or there exists b such that $(b/a)\,A \in V$.

A cds V is said to be $(\forall, A, a)$-*valuating* if $\forall x\,[(x/a)\,A] \in V$ or there exists b such that $(b/a)\,A \notin V$.

Remark 1 x is any variable such that $A\,\rangle x\langle$, but these definitions are independent of the specific variable chosen: if $A\,\rangle y\langle$, we know that

$$\exists x\,[(x/a)\,A] \equiv \exists y\,[(y/a)\,A] \quad \text{and} \quad \forall x\,[(x/a)\,A] \equiv \forall y\,[(y/a)\,A].$$

Remark 2 There exist $(\exists, A, a)$-valuating or $(\forall, A, a)$-valuating cds: in fact, any truth set has both these properties (for any A and a).

Remark 3 If $a \notin I_A$, then any cds is both $(\exists, A, a)$-valuating and $(\forall, A, a)$-valuating, for then

$$\exists x\,[(x/a)\,A] \equiv A \equiv \forall x\,[(x/a)\,A] \quad \text{and} \quad (b/a)\,A = A.$$

Remark 4 A cds may be $(\exists, A, a)$-valuating, and not $(\forall, A, a)$-valuating. However, any cds which is not $(\exists, A, a)$-valuating must be $(\forall, A, a)$-valuating, since then, for any b, we have $(b/a)\,A \notin V$.

1 PROPOSITION *A cds is* $(\forall, A, a)$-*valuating if and only if it is* $(\exists, \neg A, a)$-*valuating.*

---

* *Translator's note* French $(\exists, A, a)$-*validant*.

Proof Suppose V is $(\forall, A, a)$-valuating, and let $\exists x\,[(x/a)\neg A] \in V$. Then $\neg\forall x\,[(x/a)\,A] \in V$ and $\forall x\,[(x/a)\,A] \notin V$, so that there exists b such that $(b/a)\,A \notin V$, or $\neg(b/a)\,A \in V$ i.e. $(b/a)\neg A \in V$. Proof of the converse is analogous.

(2) We shall call a cds *(A, a)-valuating* if it is both $(\exists, A, a)$-valuating and $(\forall, A, a)$-valuating.

2 Proposition *V is (A, a)-valuating if and only if it has at least one of the following three properties:*

(1) $\exists x\,[(x/a)\,A] \notin V$;

(2) $\forall x\,[(x/a)\,A] \in V$;

(3) there exists b' such that $(b'/a)\,A \in V$ and there exists b'' such that $(b''/a)\,A \notin V$.

Proof Suppose first that V is (A, a)-valuating; then if it does not possess properties (1) and (2), it follows from the definition that it must possess property (3).

Now suppose that V has one of these three properties.

If $\exists x\,[(x/a)\,A] \in V$, so that V does not have property (1), then if it has property (2) it follows that $(b/a)\,A \in V$ for all b; if it has property (3) there exists b' such that $(b'/a)\,A \in V$; thus, in either case, V is $(\exists, A, a)$-valuating.

If $\forall x\,[(x/a)\,A] \notin V$, so that V does not have property (2), then if it has property (1) it follows that $(b/a)\,A \notin V$ for all b; if it has property (3) there exists b'' such that $(b''/a)\,A \notin V$; thus, in either case, V is $(\forall, A, a)$-valuating.

(3) Given a sentence A, we shall say that a cds V is A-valuating if it is (A, a)-valuating for all $a \in I_A$ (or, equivalently, for any individual a).

3 Proposition *The following two assertions are equivalent:*

(1) V is a truth set;

(2) V is a cds which is A-valuating for every sentence A.

Proof If V is a truth set, consider a sentence A and an individual a. If $\exists x\,[(x/a)\,A] \notin V$, we are done. Otherwise, $\exists x\,[(x/a)\,A] \in V$, and then there exists b such that $(b/x)\,(x/a)\,A = (b/a)\,A \in V$. Again, if $\forall x\,[(x/a)\,A] \in V$, the proof is complete; otherwise, there exists b such that $(b/x)\,(x/a)\,A = (b/a)\,A \notin V$. Thus V is (A, a)-valuating.

Conversely, suppose V is A-valuating for every sentence A. Let f^x be a quantifiable formula, of the form $f^x = (x/a)\,A$. Then, by assumption, if $\exists x\,[f^x] \in V$, there exists b such that $(b/a)\,A = (b/x) f^x \in V$. Thus V is a truth set.

(4) Given a subset J of E, we shall say that a cds V is J-valuating if it is A-valuating for every $A \in J$.

A truth set is thus an E-valuating cds.

Recall that if A is a sentence then S_A denotes the set of all *subsentences* of A (cf. definition on p. 18). More generally, if $J \subset E$ we set $S_J = \bigcup_{A \in I} S_A$.

The following proposition is fundamental.

4 PROPOSITION *Let $J \subset E$ and let V be an S_J-valuating cds. Then there exists a unique truth set V_h such that $V_h \cap S_J = V \cap S_J$.*

Proof Let H be the characteristic function of V, h the restriction of H to $\mathfrak{a}$, and $\tilde{h}$ the (unique) extension of h to E, which defines a truth set V_h.

We shall prove by induction on the order of $B \in S_J$ that $H(B) = \tilde{h}(B)$.

If $B \in \mathfrak{a}$, then $H(B) = h(B) = \tilde{h}(B)$.

If $B = \neg C$, then $C \in S_J$, so that by the induction hypothesis $H(C) = \tilde{h}(C)$; hence $H(B) = \neg H(C) = \neg \tilde{h}(C) = \tilde{h}(B)$.

The argument for $B = CkD$ is similar.

If $B = \exists x\, [f^x]$, then, for any a, $(a/x) f^x \in S_J$ and therefore

$$H((a/x) f^x) = \tilde{h}((a/x) f^x);$$

but V is S_J-valuating, and therefore

$$H(B) = 1 \Leftrightarrow \text{there exists } a \text{ such that } H((a/x) f^x) = 1$$

$$\Leftrightarrow \text{there exists } a \text{ such that } \tilde{h}((a/x) f^x) = 1, \text{ i.e. } \tilde{h}(B) = 1.$$

Thus $H(B) = \tilde{h}(B)$.

COROLLARY *The following two assertions are equivalent:*

(1) J is contained in a truth set [in every truth set].

(2) J is contained in an S_J-valuating cds [in every S_J-valuating cds].

For the proof of the classical completeness theorem, it would suffice to retain the particular case in which J contains a single sentence A, i.e., to consider the equivalent statements

(1) A is universally valid;

(2) A belongs to every S_A-valuating cds.

However, we shall need the general proposition for the Löwenheim-Skolem Theorem.

Transition to the Boolean ring E/R

Let U be an ultrafilter of E/R, $U = \varphi(V)$ and $V = \varphi^{-1}(U)$, where V is a cds. Let $\alpha \in E/R$, and let A' and A'' be two representatives of α. Let a be an individual.

Then V is $(\exists, A', a)$-valuating if and only if it is $(\exists, A'', a)$-valuating. The proof is immediate:

$$\exists x\,[(x/a)\,A'] \equiv [(x/a)\,A''] \quad \text{and} \quad (b/a)\,A' \equiv (b/a)\,A''.$$

This makes it possible to define a $(\exists, \alpha, a)$-valuating ultrafilter U as an ultrafilter such that the cds $V = \varphi^{-1}(U)$ is $(\exists, A, a)$-valuating for any representative A of α.

Note that an equivalent, direct definition would be:

Either $\exists a\alpha \notin U$, or there exists b such that $(b/a)\,\alpha \in U$.

We define a $(\forall, \alpha, a)$-valuating ultrafilter in analogous fashion. An ultrafilter is said to be (α, a)-valuating if it is both $(\forall, \alpha, a)$-valuating and $(\exists, \alpha, a)$-valuating, or, equivalently, $V = \varphi^{-1}(U)$ is an (A, a)-valuating cds for any representative A of α.

We call an ultrafilter α-valuating if it is (α, a)-valuating for every $a \in I_\alpha$ (we know that I_α is a finite set). This is equivalent to the statement: $V = \varphi^{-1}(U)$ is an A-valuating cds for any representative A of α.

We introduce the following notation:

$\mathscr{U}_{\exists\alpha a}$: the set of $(\exists, \alpha, a)$-valuating ultrafilters;

$\mathscr{U}_{\forall\alpha a}$: the set of $(\forall, \alpha, a)$-valuating ultrafilters;

$\mathscr{U}_{\alpha,a} = \mathscr{U}_{\exists\alpha a} \cap \mathscr{U}_{\forall\alpha a}$: the set of (α, a)-valuating ultrafilters;

$\mathscr{U}_\alpha$: the set of α-valuating ultrafilters.

Note the following formulas:

$$\mathscr{U}_{\forall\alpha a} = \mathscr{U}_{\exists\neg\alpha a}; \quad \mathscr{U}_\alpha \bigcap_{a \in I_\alpha} \mathscr{U}_{\alpha a} \quad \text{(finite intersection)}$$

$$\mathscr{U}_{\exists\alpha a} = \sigma\,(\neg\exists a\alpha) \cup \bigcup_b \sigma\,((b/a)\,\alpha).$$

2 Topological properties (Baire spaces)

Let X be the dual space of E/R—the set of ultrafilters. We shall now make repeated use of an important theorem of general topology: every (locally) compact topological space, *a fortiori* every Boolean space, is a *Baire space*, that is to say, every denumerable intersection of everywhere dense open sets is an everywhere dense set. However, it is easy to prove this theorem in the particular case of Boolean spaces:

LEMMA *Let X be a Boolean space, and let* $\{\Omega_n\}_{n=1}^{\infty}$ *be a denumerable family of everywhere dense open sets. Then* $\bigcap \Omega_n$ *is everywhere dense.*

Proof Let $\sigma(\beta)$ be any nonempty clopen set. Then $\sigma(\beta) \cap \Omega_1$ is a nonempty open set, and so there exists a clopen set $\sigma(\beta_1)$ such that

$$\sigma(\beta) \cap \Omega_1 \supset \sigma(\beta_1), \quad \sigma(\beta_1) \neq \emptyset.$$

Similarly, there exists a clopen set $\sigma(\beta_2)$ such that

$$\sigma(\beta_1) \cap \Omega_2 \supset \sigma(\beta_2), \quad \sigma(\beta_2) \neq \emptyset;$$

and so on.

Now $\sigma(\beta_n)$ is a decreasing sequence of nonempty closed sets; since the space is compact, it follows that

$$\bigcap \sigma(\beta_n) \neq \emptyset.$$

Now for any n, $\sigma(\beta_n) \subset \Omega_h$, and so $\bigcap \sigma(\beta_n) \subset \bigcap \Omega_n$. But, on the other hand, $\sigma(\beta) \supset \sigma(\beta_1) \supset \sigma(\beta_2) \cdots$, so that $\bigcap \sigma(\beta_n) \subset \sigma(\beta)$. Thus $\sigma(\beta) \cap \bigcap \Omega_n \supset \sigma(\beta_n) \neq \emptyset$, and this implies that $\bigcap \Omega_n$ is everywhere dense.

We now have the following properties.

5 PROPOSITION *Every set* $\mathscr{U}_{\exists\alpha a}$ *is an open set.*

For, by the preceding formula, it is a union of clopen sets.

6 PROPOSITION *Every set* $\mathscr{U}_{\forall\alpha a}$ *is an open set.*

7 PROPOSITION *Every set* $\mathscr{U}_{\exists\alpha a}$ *is dense.*

Proof Let U be any ultrafilter which is not a limit point of $\mathscr{U}_{\exists\alpha a}$. Then $U \notin \mathscr{U}_{\exists\alpha a}$, so that $\exists a\alpha \in U$; also, there is a clopen set $\sigma(\beta)$ such that $U \in \sigma(\beta)$ and $\sigma(\beta) \subset \complement\mathscr{U}_{\exists\alpha a}$.

Let b be any individual, and consider $(b/a)\,\alpha$:

If $(b/a)\,\alpha \neq 0$, let U' be an ultrafilter containing $(b/a)\,\alpha$. Then U' contains $\exists a\alpha$, and U' is $(\exists, \alpha, a)$-valuating. Therefore U' does not contain β, and so $\neg\beta \in U'$.

Therefore $U' \in \sigma(\neg\beta)$, and it follows that $\sigma((b/a)\,\alpha) \subset \sigma(\neg\beta)$; this is *a fortiori* true when $(b/a)\,\alpha = 0$.

Hence $(b/a)\,\alpha \leqslant \neg\beta$ for all b, and therefore $\exists a\alpha \leqslant \neg\beta$.

Now $\exists a\alpha \in U$, so that $\neg\beta \in U$; but this is a contradiction, since $\beta \in U$.

8 PROPOSITION *Every set* $\mathscr{U}_{\forall\alpha a}$ *is everywhere dense.*

9 PROPOSITION *Every set* $\mathscr{U}_{\alpha a}$ *is an everywhere dense open set.*

In fact, it is the intersection of two open sets, therefore open, and the lemma is applicable.

10 PROPOSITION *Every set $\mathscr{U}_\alpha$ is an everywhere open set.*

For it is a finite intersection of open sets, therefore open, and the lemma is applicable.

These properties will enable us to prove the semantic completeness theorem, first in a special case, and then in all its generality.

3 Semantic completeness theorem

Special case: the set I of individuals is denumerable

In this case it is easy to see that, for any sentence A, the set S_A is denumerable. This follows by induction on the order of A:

If A is an atomic sentence, then $S_A = \{A\}$.

If $A = \neg B$, then $S_A = S_B \cup \{A\}$, and since S_B is denumerable by the induction hypothesis, so is S_A.

If $A = BkC$, then $S_A = S_B \cup S_C \cup \{A\}$, and since S_B and S_C are denumerable by hypothesis, so is S_A.

If $A = Qx\,[f^x]$, then $S_A = \bigcup_a S_{(a/x)f^x} \cup \{A\}$ is the union of denumerably many denumerable sets.

Now let A be a sentence that belongs to every truth set, i.e., a universally valid sentence.

As we have already seen, A belongs to every S_A-valuating cds.

Set $S_A = \{A_n\}_N$, $\alpha = \varphi(A)$, $\alpha_n = \varphi(A_n)$.

Let $A \notin T$. Then $\alpha \neq 1$, therefore $\neg\alpha \neq 0$. Thus $\sigma(\neg\alpha)$ is a nonempty clopen set. Now every $\mathscr{U}_{\alpha_n}$ is an everywhere dense open set, and so $\bigcap \mathscr{U}_{\alpha_n}$ is everywhere dense (the Baire property—see above). Thus $\sigma(\neg\alpha) \cap \bigcap \mathscr{U}_{\alpha_n} \neq \emptyset$ and there exists an ultrafilter U such that 1) $\neg\alpha \in U$, and 2) U is α_n-valuating for any n. Let $V = \varphi^{-1}(U)$. Then $\neg A \in V$, and V is A_n-valuating for all n, i.e., S_A-valuating. But $A \in V$, and this is a contradiction.

Thus $A \in T$.

Thus, when I is denumerable, every universally valid sentence is a provable sentence.

General case

We now resume the general case, in which I is any infinite set.

Let A be a universally valid sentence of E, and I^* a denumerable subset of I such that $I_A \subset I^*$.

Then A also belongs to the sublogic E^* based on I^*.

It is easily seen that A is universally valid in E^*: For let s be the simultaneous substitution defined by

$$s(a) = \begin{cases} a & \text{if } a \in I^*, \\ a_0 & \text{if } a \notin I^*, \end{cases}$$

where a_0 is any fixed individual of I^*.

s maps I onto I^*, and therefore (cf. Proposition 40 of Chapter III) if V^* is a truth set of E^*, then $s^{-1}V^*$ is a truth set of E. Therefore $A \in s^{-1}V^*$, or $s(A) \in V^*$. But $s(A) = A$, and so $A \in V^*$.

Thus, from the denumerable case, treated above, it follows that $A \in T^*$.

But $T^* = T \cap E(I^*)$, and therefore $A \in T$.

To summarize, we can thus state, for the predicate calculus:

SEMANTIC COMPLETENESS THEOREM *Every universally valid sentence is a provable sentence.*

Remark This theorem is the exact converse of the Semantic Consistency Theorem. Combining the two theorems, we see, as in the propositional calculus, that the following two assertions are equivalent:

(1) A is a universally valid sentence ($\models A$).
(2) A is a provable sentence ($\vdash A$).

4 Various classical forms of the theorem

There are various forms of the Semantic Completeness Theorem just proved; these are associated essentially with the names of Gödel, Löwnheim and Skolem. We first introduce a few definitions.

A *domain of individuals* is any infinite subset I^* of I.

A sentence $A \in E$ is said to be *universally valid in a domain* I^* if 1) $I_A \subset I^*$, 2) A belongs to every truth set of $E^* = E(I^*)$.

Given a sentence $A \langle a_1, \ldots, a_n \rangle$ and a domain I^*, suppose that

$$\{a_1, \ldots, a_p\} \subset I - I^*,$$
$$\{a_{p+1}, \ldots, a_n\} \subset I^*.$$

Chose p distinct individuals $b_1, \ldots, b_p$ of I^*, different from $a_{p+1}, \ldots, a_n$. Then we shall call the sentence

$$A^* = (b_1/a) \cdots (b_p/a_p)\, A$$

a *model** of A in I^*. Note that $A^* \in E^*$ and $A = (a_1/b_1) \cdots (a_p/b_p)\, A^*$.

* *Translator's note* The term "model" is used in a completely different sense by English-speaking authors.

11 PROPOSITION *The following assertions are equivalent:*

(1) A is a provable sentence.
(2) A is universally valid (in I).
(3) A is universally valid in every domain I^*.
(4) A is universally valid in some domain I^*.
(5) A has a model which is universally valid in every domain I^*.
(6) A has a model which is universally valid in some domain I^*.

Proof $1 \Rightarrow 2$: This is the consistency theorem.

$2 \Rightarrow 3$: If A is universally valid, then $A \in T$, and if moreover $A \in E(I^*)$, then $A \in T^*$; thus A is universally valid in I^*, by the completeness theorem.

$3 \Rightarrow 4$: trivial.

$4 \Rightarrow 1$: this follows from the fact that $A \in T^*$, and so $A \in T$.

As for (5) and (6), it suffices to remark that if A^* is a model of A in I^*, then $A \in T$ if and only if $A^* \in T^*$.

In particular, we have the following two classical theorems:

GÖDEL'S THEOREM *If A is universally valid (or has a universally valid model) in a denumerable domain, then A is a provable sentence.*

LÖWENHEIM'S THEOREM *If A is universally valid in a denumerable domain, then A is universally valid in any domain.*

We now consider some more definitions.

A sentence $A \in E$ is said to be *satisfiable* if there exists at least one system of truth values h such that $\tilde{h}(A) = 1$, i.e., if A belongs to at least one truth set. Note that this is equivalent to the statement: A is consistent (i.e. $\neg A \notin T$). The proof is easy:

If A is satisfiable, then there exists h such that $\tilde{h}(\neg A) = 0$, so that $\neg A \notin T$.

If A is consistent, $\neg A \notin T$; then $\neg A$ is not universally valid, and therefore there is at least one system of truth values h such that $\tilde{h}(\neg A) = 0$, i.e. $\tilde{h}(A) = 1$.

Let us call a sentence $A \in E$ *satisfiable in a domain* I^* if (1) $I_A \subset I^*$, (2) A belongs to at least one truth set of E^*.

12 PROPOSITION *The following statements are equivalent:*

(1) A is satisfiable.
(2) A is satisfiable in some domain I^*.
(3) A is satisfiable in every domain I^*.

To prove this we need only transform the statement $\neg A \notin T$ means of the previous equivalences.

Remark 1 The intersection of all truth sets of E is T, or, what is the same, the intersection of all valuating ultrafilters of E/R is $\{1\}$.

Remark 2 The set $\mathscr{U}$ of valuating ultrafilters is dense in X. For let $\sigma(\alpha)$ be a nonempty clopen set. Then $\alpha \neq 0$, and therefore α belongs to at least one valuating ultrafilter. Thus $\sigma(\alpha)$ contains at least one valuating ultrafilter.

As in the case of the propositional calculus (p. 98), we have a one-to-one mapping of $\mathscr{U}$ onto $\mathbb{U}^a$: $\Gamma(x) = h$, where x is a valuating ultrafilter and h the system of truth-values such that $\bar{h}$ is the characteristic function of x. Γ is again a continuous mapping of $\mathscr{U}$ (with the topology induced by that of X) onto the Cantor space $\mathbb{U}^a$. For if $V = \prod \Omega_u$ is a nonempty elementary set of $\mathbb{U}^a$ (not coinciding with the entire space), then, as for L':

$$\Gamma^{-1}(V) = \sigma(\alpha) \cap \mathscr{U},$$

and the latter is a clopen set of $\mathscr{U}$, where $\alpha = \varphi(A_1 \wedge \cdots \wedge A_n)$:

$$A_i = \begin{cases} u_i & \text{if } \Omega_{u_i} = \{1\} \\ \neg u_i & \text{if } \Omega_{u_i} = \{0\}. \end{cases}$$

However, in this case Γ is not a homeomorphism, i.e. Γ^{-1} is not continuous, since otherwise $\mathscr{U}$ would be compact, therefore closed in X, and therefore $\mathscr{U} = X$, which is false.

It is also possible to show directly that Γ^{-1} is discontinuous at certain points (thus proving, in a new way, that $\mathscr{U}$ is not compact, i.e. $\mathscr{U} \neq X$). To do this, consider a predicate r_n^p of nonzero weight, and $p-1$ fixed individuals $b_2, \ldots, b_p$. Let h_0 be a system of truth values such that $h_0(r_n^p ab_2 \cdots b_p) = 1$ for any $a \in I$.

h_0 defines a valuating ultrafilter x_0, associated with a truth set V_0. Let $A = \forall x\, [r_n^p xb_2 \cdots b_p]$; then $A \in V_0$. Set $\alpha = \varphi(A)$, and so $\alpha \in x_0$, and therefore $\sigma(\alpha) \cap \mathscr{U}$ is a neighborhood of x_0 in $\mathscr{U}$, and $h_0 = \Gamma(x_0)$ or $x_0 = \Gamma^{-1}(h_0)$.

On the other hand, consider any neighborhood of h_0, i.e. an elementary set $W = \prod \Omega_u$ in $\mathbb{U}^a$ such that $h_0(u) \in \Omega_u$ for all u. $\Omega_u = \mathbb{U}$ for all but a finite number of indices. Thus there exists an individual a_1 such that $\Omega_{u_1} = \mathbb{U}$, where $u_1' = r_n^p a_1 b_2 \cdots b_p$. Therefore, there exists $h \in W$ such that $h(u_1) = 0$. Then $\bar{h}(A) = 0$, and thus, if $x = \Gamma^{-1}(h)$, we have $x \notin \sigma(\alpha) \cap \mathscr{U}$. This proves that Γ^{-1} is discontinuous at h_0.

Remark 3 The semantic completeness theorem has the following equivalent topological formulation: $\mathscr{U}$ is dense in X.

For, assuming the statement true, consider a universally valid sentence A and let $\alpha = \varphi(A)$. If $\alpha \neq 1$, then $\neg\alpha \neq 0$, and so $\sigma(\neg\alpha) \cap \mathscr{U} \neq \emptyset$, which is a contradiction.

Syntactic incompleteness

We have just proved that the same semantic completeness theorem holds true for the predicate calculus as for the propositional calculus. By contrast, here we cannot prove any analogue of the syntactic completeness theorems. In fact, one can even prove that the predicate calculus is not complete in the syntactic sense, in other words:

One can add a new and independent axiom schema to the list of axiom schemata, in such a way that the new calculus is not contradictory.

To see this, consider the axiom schema

$$\exists \bar{x}\, [f^{\bar{x}}] \to \forall \bar{x}\, [f^{\bar{x}}],$$

and let J be the set of all instances of this schema.

(a) To show that the schema is independent, consider the instance

$$A = \exists x\, [r_n^p\, x b_2 \cdots b_p] \to \forall x\, [r_n^p\, x b_2 \cdots b_p],$$

where r_n^p is any predicate of nonzero weight.

Let a_1 and a_1' be two distinct individuals, and h a system of truth values such that

$$h\,(r_n^p\, a_1 b_2 \cdots b_p) = 1,$$
$$h\,(r_n^p\, a_1' b_2 \cdots b_p) = 0.$$

Then $\tilde{h}(A) = 0$, so that $A \notin T$.

(b) J is consistent, since for any uniform system of truth values (p. 33) we have

$$\tilde{h}(A) = 1 \quad \text{for all } A \in J, \text{ so that } J \subset V_h.$$

Note that, though it is not contradictory, a predicate calculus containing this axiom schema is rather uninteresting, since to all intents and purposes it reduces to the propositional calculus.

Remark From the point of view of the quotient ring E/R, the previous arguments have the following result:

For any $\alpha \in E/R$ and any $a \in I$:

$$\exists a\alpha \to \forall a\alpha \neq 0$$

or $$\neg\exists a\alpha \vee \forall a\alpha \neq 0.$$

In other words, it is impossible that

$$\exists a\alpha = 1 \quad \text{and} \quad \forall a\alpha = 0$$

simultaneously, or, it is impossible that

$$\exists a\alpha = \exists a(\neg\alpha) = 1.$$

Another result is that we can never have

$$\exists a\alpha = \exists a(\neg\alpha)$$

since otherwise we should have $\exists a\alpha = \exists a\alpha \vee \forall a(\neg\alpha) = \exists a(\alpha \vee \neg\alpha) = 1$, which brings us back to the previous case.

5 The Löwenheim-Skolem theorem

Recall the definition of satisfiability of a sentence A: A is satisfiable if it belongs to at least one truth set. We have seen that this is equivalent to the consistency of A (i.e. $\neg A \notin T$). We now propose to generalize this assertion to arbitrary sets J of sentences. An obvious difficulty is that one cannot hope to prove that any consistent subset is contained in a truth set. Were this true, any cds would be contained in a truth set, and would thus be a truth set. We shall therefore generalize the concept of satisfiability to sets of sentences.

Let I^* be a domain of individuals (either contained in or containing I), and E^* the logic based on I^* (which is either a sublogic or a superlogic of E). Let $J \subset E$. Then:

(1) J is said to be *directly satisfiable* in I^* if a) $I_J \subset I^*$ and b) there exists a truth set of E^* that contains J.

(2) J is said to be *satisfiable* in I^* if there exists a mapping $s\colon I \hookrightarrow I^*$ whose restriction to I_J is one-to-one, such that $s(J)$ is contained in a truth set of E^*. We shall then call $s(J)$ a *model* of J in I^*, and this model is directly satisfiable in I^*.

LEMMA *Let E^+ be the superlogic of E based on the domain $I^+ = I \cup \{\varepsilon\}$, where ε is a new symbol. Let V be a cds of E and f^x a quantifiable formula.*

(1) If $\exists x\,[f^x] \in V$, but for all $a \in I$, $(a/x)f^x \notin V$, then there exists a cds V^+ in E^+ such that $V, (\varepsilon/x)f^x \subset V^+$.

(2) If $\forall x\,[f^x] \notin V$, but for all $a \in I$, $(a/x)f^x \in V$, then there exists a cds V^+ in E^+ such that $V, \neg(\varepsilon/x)f^x \subset V^+$.

Proof We prove (1)—the proof of (2) is quite similar.

Suppose the assertion is false, i.e. V, $(\varepsilon/x)f^x$ are inconsistent in E^+. Then there exists a finite subset V^0 of V such that V^0, $(\varepsilon/x)f^x$ is inconsistent. Then, in E^+:

$$V^0, (\varepsilon/x)f^x \vdash \omega, \quad \text{or} \quad V^0 \vdash \neg(\varepsilon/x)f^x, \quad \text{or} \quad \wedge V^0 \to \neg(\varepsilon/x)f^x \in T^+.$$

Thus, by a rule of inference,

$$\wedge V^0 \to \forall x\,[\neg f^x] \in T^+,$$

and hence $\qquad \wedge V^0 \to \forall x\,[\neg f^x] \in T,\quad$ since $T = T^+ \cap E$.

Thus, in E: $\quad V^0 \vdash \neg\exists x f^x$, which contradicts $\exists x\,[f^x] \in V$.

COROLLARY *Let A be any sentence and V a cds of E. If V is not A-valuating, then one can add to I a finite number of new individuals in such a way that V is contained in an A-valuating cds V^+ of the resulting superlogic E^+.*

1 THEOREM *Let J be a denumerable and consistent subset of E such that $\complement I_J$ is infinite. Then J is directly satisfiable in a denumerable subdomain I^* of I.*

Proof Since J is denumerable, so is I_J. But since I_J may be finite, let us consider an infinite denumerable domain I_0 (a subset of I) containing I_J, such that $\complement I_0$ is infinite. J is also a consistent and denumerable subset of the sublogic E_0 based on I_0; thus J is contained in a cds V_0 of E_0. Partition the infinite set $\complement I_0$ into two infinite disjoint subsets K_0 and K_0'. Let $S_J(E_0)$ denote the set of subsentences of sentences of J in the logic E_0; this set is denumerable.

(1) Applying the preceding Corollary to all sentences B_n ($n \in N$) of $S_J(E_0)$ in succession, we can construct a denumerable chain of sublogics E_n' of E_0 (which are of course sublogics of E), such that for every n:

(a) the domain I_n' is denumerable, $I_n' - I_{n-1}'$ is finite, and $I_{n-1}' \subset I_n' \subset I_0 \cup K_0$;

(b) there exists a cds V_n' of E_n' which contains V_{n-1}' and is $\{B_1, B_2, \ldots, B_n\}$-valuating.

Let E_1 be the logic based on $I_1 = \bigcup_{n\in N} I_n'$.

As the denumerable union of denumerable sets, the set I_1 is denumerable. $\complement I_1$ is infinite, since $\complement I_1$ contains the infinite set K_0'; $I_0 \subset I_1 \subset I$.

We first show that the set T_1 of provable sentences of E_1 is $\bigcup_{n\in N} T_n'$ (where T_n' is the set of provable sentences of E_n').

Indeed, if $A \in T_1$ there exists n such that $A \in E'_n$, so that

$$A \in T_1 \cap E'_n = T'_n \Rightarrow A \in \bigcup_n T'_n.$$

Conversely, if $A \in \bigcup_n T'_n$ there exists n such that $A \in T'_n \subset T_1 \Rightarrow A \in T_1$.

We now claim that the set $V_1 = \bigcup_n V'_n$ is a cds in E_1. In fact:

$T_1 \subset V_1$, since $T'_n \subset V'_n$ for every n.

If A, $A \to B \in V_1$, then there exists n such that A, $A \to B \in V'_n$ (since the cds V'_n form a chain), and thus $B \in V'_n \Rightarrow B \in V_1$.

Finally, given a sentence $A \in E_1$, we must show that either $A \in V_1$ or $\neg A \in V_1$. Indeed, if $A \notin V_1$, then $A \notin V'_n$ for every n. But there exists p such that $A \in E_p$. Thus $A \notin V'_p \Rightarrow \neg A \in V'_p \Rightarrow \neg A \in V_1$.

To summarize: we have constructed a superlogic E_1 of E_0 with the following properties:

(a) I_1 is denumerable, $I_0 \subset I_1 \subset I$, and $\complement I_1$ is infinite;

(b) there exists a cds V_1 of E_1 which contains V_0 and is $S_J(E_0)$-valuating.

Note that V_1 need not be $S_J(E_1)$-valuating.

(2) Utilizing the results of the previous section, we continue the construction, constructing a denumerable chain of superlogics E_n of E_0 with the following properties:

(a) I_n is denumerable, $I_{n-1} \subset I_n \subset I$, and $\complement I_n$ is infinite;

(b) there exists a cds V_n of E_n which contains V_{n-1} and is $S_J(E_{n-1})$-valuating.

Now let E^* be the superlogic of E_0 based on $I^* = \bigcup_n I_n$. *I^* is denumerable, contains I_0 (therefore also I_J), and is contained in I.*

Let $V^* = \bigcup_n V_n$. The proof of the preceding paragraphs shows that *V^* is a cds in E^* containing J.*

We now claim that V^* is $S_J(E^*)$-valuating. In fact, if $B \in S_J(E^*)$, then there exists $n \in N$ such that $B \in S_J(E_n)$. The cds V_{n+1} of E_{n+1} is B-valuating, and so, *a fortiori*, is V^*.

Thus J is contained in an $S_J(E^*)$-valuating cds V^* of E^*. By the Corollary to Proposition 4, this means that J is contained in a truth set of E^*, which completes the proof of Theorem 1.

2 Theorem *If J is a denumerable consistent subset of E, then J is satisfiable in a denumerable subdomain I^* of I.*

(Here we do not assume that $\complement I_J$ is infinite.)

Proof If $\complement I_J$ is infinite, Theorem 1 is applicable.

If $\complement I_J$ is finite, it follows that I itself must be infinite and denumerable, therefore equipollent to N. Therefore, there exists a one-to-one mapping f of I onto N.

I_J is also infinite denumerable, therefore equipollent to the set of even integers $\mathscr{P}$. So there exists a one-to-one mapping g of I_J onto $\mathscr{P}$.

$I' = f^{-1}(\mathscr{P})$ is an infinite subset of I, and $\complement I' = f^{-1}(\mathscr{I})$ is also infinite (where $\mathscr{I}$ is the set of odd integers).

Now define a mapping s: $I \hookrightarrow I'$ by

$$s(a) = \begin{cases} f^{-1}(g(a)) & \text{if } a \in I_J; \\ f^{-1}(0) & \text{if } a \notin I_J. \end{cases}$$

For any $b \in I'$, we have $f(b) \in \mathscr{P}$; therefore there exists a unique $a \in I_J$ such that $f(b) = g(a)$, so that $b = s(a)$. This shows that s maps I_J one-to-one onto $I' = s(I_J)$.

Now $s(J)$ is a denumerable and consistent subset of E. $I_{s(J)} = s(I_J)$, so that $\complement I_{s(J)} = \complement I'$ is infinite, and Theorem 1 is applicable. Thus $s(J)$ is directly satisfiable in a denumerable subdomain I^*, and J is satisfiable in I^*.

A few results

Many theorems may be derived as corollaries of the theorem just proved. We collect them in the following proposition.

13 PROPOSITION *Let J be a denumerable subset of E. Then the following assertions are equivalent:*

(1) J is consistent.
(2) J is satisfiable in a denumerable domain.
(3) J is satisfiable in a (not necessarily denumerable) domain.
(4) Every finite subset of J is directly satisfiable in some domain.

Proof $1 \Rightarrow 2$: This is Theorem 2.

$2 \Rightarrow 3$: Trivial, since every denumerable set is equipollent to a subset of an infinite set.

$3 \Rightarrow 1$: Suppose that J is satisfiable in a domain I^*. Then there is a mapping s of I into I^* such that $s(J)$ is contained in a truth set V^* of E^*. Thus J is contained in $s^{-1}(V^*)$, which is a cds of E (cf. p. 59): J is consistent.

$1 \Leftrightarrow 4$: Trivial, in view of the equivalence $1 \Leftrightarrow 3$, since J is consistent if and only if every finite subset of J is consistent.

We recall the classical forms of these theorems:

LÖWENHEIM-SKOLEM THEOREM *If a denumerable set of sentence is satisfiable in some domain, then it is satisfiable in a denumerable domain.*

COMPACTNESS THEOREM *A sufficient condition for a denumerable set of sentences J to be satisfiable in some domain is that every finite subset of J have this property.*

Generalization

Let J be a consistent subset of E, of *arbitrary* cardinality. We can now prove the following result.

3 THEOREM *There exists a superlogic* E^+ *of* E, *based on a domain of individuals* I^+, *such that*

(a) J *is directly satisfiable in* E^+;
(b) card $(I^+) \leqslant$ max (card (J), card (I)).

Proof Note first that card $(S_J(E)) \leqslant$ max (card (I), card (J)) $=$ card (I) . card (J).

Well-order the set $S_J(E)$, and let B_0 be its smallest element, where we use the notation $S_J(E) = \{B_\gamma\}$. Since J is consistent, it is contained in a cds V in E.

By the Corollary on p. 112, one can add a finite number of individuals to I (obtaining a superdomain I_0), in such a way that V is contained in a B_0-valuating cds V_0 in the corresponding superlogic E_0.

We now prove the following property $P(\gamma)$ by transfinite induction:

There exist a superlogic E_γ of E, based on a domain I_γ, and a cds V_γ in E_γ which is B_ω-valuating for all $\omega \leqslant \gamma$, such that for all $\omega \leqslant \gamma$: $I_\omega \subset I_\gamma$ and $V_\omega \subset V_\gamma$.

$P(0)$ is true, as we have just shown.

Assume that $P(\gamma)$ is true for $0 \leqslant \gamma < \delta$, and let us prove $P(\delta)$.

Let E' be the superlogic based on $I' = \bigcup_{0 \leqq \gamma < \delta} I_\gamma$, and let $V' = \bigcup_{0 \leqq \gamma < \delta} V_\gamma$ (which is a cds in E'). V' is B_γ-valuating for every $\gamma < \delta$. By adding at most finitely many elements to I', we can construct a new superlogic E'', based on I'', for which there exists a B_δ-valuating cds $V'' \supset V'$. It then suffices to take $E_\delta = E''$, $I_\delta = I''$, and $V_\delta = V''$.

Now let $I^1 = \bigcup I_\gamma$, and let E^1 be the superlogic based in I^1; $V^1 = \bigcup V_\gamma$ is an $S_J(E^1)$-valuating cds in E^1, and $J \subset V^1$.

Since V^1 is not necessarily $S_J(E^1)$-valuating we resume the above process, and repeat it a denumerable number of times: there exist E^2 based on $I^2 \supset I^1$ and a cds $V^2 \supset V^1$ which is $S_J(E^1)$-valuating, and so on.

Finally, we obtain a superlogic E^+ based on $I^+ = \bigcup I^n$ and a cds $V^+ = \bigcup V^n$ which is $S_J(E^+)$-valuating. It follows that J is contained in a truth set of E^+.

Moreover:

$$\text{card}\,(I^1) \leqslant \sum_\gamma \text{card}\,(I_\gamma) \leqslant \max\,(\text{card}\,(I), \text{card}\,(J)),$$

and

$$\text{card}\,(I^+) \leqslant \sum_n \text{card}\,(I^n) \leqslant \max\,(\text{card}\,(I), \text{card}\,(J)).$$

Remark It follows, in particular, that every cds of E is contained in a truth set of some superlogic E^+, in other words, every cds of E is obtained by restriction of a "super-truth-set".

CHAPTER VIII

Predicate calculus with equality

THE CALCULUS L'' THAT we have defined is completely general in nature, in that we have attributed no meaning to the predicates; provable sentences correspond to true assertions, independently of the possible sense of the relations represented by the predicates appearing therein. We now describe a more specific version of this logical calculus, by singling out certain predicates to represent well-known relations; of course, this will necessitate the introduction of appropriate new axioms.

In this chapter we propose to describe the first-order predicate calculus with equality, i.e., with a distinguished binary predicate and the relevant axioms. We shall prove that the fundamental consistency and completeness theorems remain valid in the new calculus.

1 The concept of *J*-equality

Consider, for the moment, the set E of sentences of the ordinary calculus L''.

Let a and b be two fixed individuals; we define a relation in E, denoted by

$$A \approx A' \bmod (a/b),$$

by $(a/b)\,A = (a/b)\,A'$.

This is clearly an equivalence relation over E. Denote the equivalence class of a sentence A by $(a/b, A)$.

The following properties are easily verified:

(*1*) $(a/a, A) = \{A\}$.

(2) If $a \notin I_A$ and $b \notin I_A$, then $(a/b, A) = \{A\}$.

(3) $(a/b)\,A \in (a/b, A)$, since $(a/b)\,A = (a/b)\,(a/b)\,A$.

(4) $(b/a)\,A \in (a/b, A)$, since $(a/b)\,(b/a)\,A = (a/b)\,A$.

(5) $(a/b, A) = (b/a)\,A)$.

For if $A' \in (a/b, A)$, then $(a/b)\,A' = (a/b)\,A$

$$(b/a)\,(a/b)\,A' = (b/a)\,(a/b)\,A$$

$$(b/a)\,A' = (b/a)\,A, \quad \text{so that} \quad A' \in (b/a, A).$$

(6) Every class $(a/b, A)$ is finite.

In fact, if a has n occurrences in $(a/b)\,A$ and $A' \in (a/b, A)$, i.e. $(a/b)\,A' = (a/b)\,A$, then A' differs from $(a/b)\,A$ only in that some or all of these

n occurrences have been replaced by occurrences of a or b, and this gives 2^n possibilities.

Now let J be a fixed subset of E, and define a relation, which we denote by $a =_J b$ and call J-equality, in the set of individuals I:

$a =_J b$ if and only if J is saturated* with respect to equivalence mod (a/b).

By property (3) above, an equivalent definition is:

$a =_J b$ if and only if, for every sentence A, $A \in J \Leftrightarrow (a/b)\, A \in J$.

1 PROPOSITION *J-equality is an equivalence relation in I.*

Proof *Reflexivity:* Property (1).

Symmetry: Property (5).

Transitivity: Assume that J is saturated mod (a/b) and mod (b/c). Let $A \in J$ and $A' \in (a/c, A)$.

$A \in J$, therefore $(b/c)\, A \in J$ and $(a/b)\,(b/c)\, A \in J$,

i.e. $(a/b)\,(a/c)\, A \in J$,

and hence $(a/b)\,(a/c)\, A' \in J$

(since $(a/c)\, A = (a/c)\, A'$).

Thus $(a/b)\,(b/c)\, A' \in J$

and so $(b/c)\, A' \in J$ and $A' \in J$.

2 Predicate calculus with equality (or calculus L''_e)

We shall employ the same symbolism as for calculus L'', but here we assume, in addition, that there is a predicate of weight 2, chosen once and for all, which we shall call "*equality*" and denote by e. Those of the atomic sentences having the form eab will be called *atomic equations*. In all other respects, sentences and formulas are defined in the usual way.

There are two additional axioms:

$$\text{S12:}\quad e\bar{a}\bar{a}$$

$$\text{S13:}\quad e\bar{a}\bar{b} \rightarrow (\bar{A} \rightarrow (\bar{b}/\bar{a})\, \bar{A}).$$

There are no additional rules of inference, and the concepts of formal proof and provable sentence are defined as for the ordinary predicate calculus.

We denote this new calculus by L''_e, and the sets of atomic sentences, sentences, and provable sentences are denoted respectively by $\mathfrak{a}$, E, and T_e.

---

* *Translator's note* I.e., J is a union of equivalence classes mod (a/b).

The calculus L''_e may always be associated with the ordinary calculus L'', based on the same alphabet, and therefore containing the predicate e. The sole difference is in the addition of the two axiom schemata above, and thus, if T denotes (as before) the set of provable sentences of L'', we have

$$T \subsetneqq T_e.$$

The concept of deducibility from hypotheses in L''_e is defined as in Chapter III (p. 39), and the notation will be $J \vdash_e A$.

It is obvious that all the propositions established for the calculus L'' remain valid. We note the following additional properties.

2 PROPOSITION $eab \vdash_e eba$.

Proof $eab \vdash_e eac \rightarrow ebc$ by S13

$eac \vdash_e eab \rightarrow ebc$ by the Deduction Theorem.

In particular: $eaa \vdash_e eab \rightarrow eba$.
But since $\emptyset \vdash_e eaa$, we have $\emptyset \vdash_e eab \rightarrow eba$.

3 PROPOSITION $eab, ebc \vdash_e eac$.

Proof $eab \vdash_e eba$

$eba \vdash_e ebc \rightarrow eac$ by S13

$eab \vdash_e ebc \rightarrow eac$.

4 PROPOSITION $eab \vdash_e A \leftrightarrow (b/a)\, A$.

Proof $eab \vdash_e A \rightarrow (b/a)\, A$

and $eab \vdash_e \neg A \rightarrow \neg (b/a)\, A$, whence $eab \vdash_e (b/a)\, A \rightarrow A$.

Deductive systems and complete deductive systems are defined in L''_e as in L''. One important property of cds V in L''_e is the following:

5 PROPOSITION $eab \in V \Leftrightarrow a =_V b$.

Proof Let $eab \in V$. If $A \in V$, then $A \rightarrow (b/a)\, A \in V$, so that $(b/a)\, A \in V$. Conversely, if $(b/a)\, A \in V$, then $A \in V$, since otherwise $\neg A \in V$ and $\neg (b/a)\, A \in V$. Thus $a =_V b$.

Now suppose that $a =_V b$. $eaa \in T_e$, therefore $eaa \in V$. Thus $(a/b)\, eab \in V$, and so $eab \in V$.

From the semantic viewpoint, it is clear that any interpretation must preserve the properties of equality. To be precise, we define:

A *system of truth-values* for the calculus L''_e is any mapping h of $\mathfrak{a}$ into $\mathbb{U}$ which satisfies the following condition for any two individuals:

$$h(eab) = 1 \Leftrightarrow h(u) = h((a/b)\,u) \quad \text{for all} \quad u \in \mathfrak{a}.$$

Setting $W_h = h^{-1}(1)$, we get the following equivalent formulation of the condition:

$$eab \in W_h \Leftrightarrow a =_{W_h} b$$

or

$$h\,(eab) = \inf_u\,[h(u) + h\,((a/b)\,u) + 1].$$

Remark 1 The condition is satisfied by the mapping which maps each element of $\mathfrak{a}$ onto 1.

Remark 2 It follows from the condition that $h(eaa) = 1$ for all $a \in I$.

A system of truth values h may be extended uniquely to a mapping $h: E \hookrightarrow \mathbb{U}$ satisfying the classical Proposition 1 of Chapter II, corresponding to the formation of sentences.

The *Semantic Consistency Theorem* remains valid:

If $A \in T_e$, then $\tilde{h}(A) = 1$ for any system of truth values of L''_e.

Proof It is sufficient to verify the assertion when A is an instance of S12 or S13.

For S12 this is trivial, for we have already seen that $h(eaa) = 1$.

To prove that $h\,(eab) \rightarrow (\tilde{h}(B) \rightarrow \tilde{h}\,((b/a)\,B) = 1$, we shall show by induction on the order of B that $h\,(eab) = 1 \Rightarrow \tilde{h}(B) = \tilde{h}\,((b/a)\,B)$.

If $B \in \mathfrak{a}$, this follows from the definition.

If $B = \neg C$ or $B = CkD$, where C and D satisfy the assertion: immediate.

Let $B = \exists x\,[f^x]$. Now $\tilde{h}(B) = 1$ if and only if there exists c such that $\tilde{h}\,((c/x) f^x) = 1$, whence, by the induction hypothesis, we get $\tilde{h}\,((b/a)(c/x)f^x) = 1$. If $c \neq a$, then $(b/a)\,(c/x) f^x = (c/x)\,(b/a) f^x$, so that $\tilde{h}\,(\exists x\,[(b/a)f^x]) = 1$. If $c = a$, then $(b/a)\,(a/x) f^x = (b/x)\,(b/a) f^x$, so that $\tilde{h}\,(\exists x\,[(b/a)f^x]) = 1$. If $B = \forall x\,[f^x]$, the proof is analogous.

The Syntactic Consistency Theorems follow similarly.

We can thus define truth sets by the equality $V_h = h^{-1}(1)$, where h is any system of truth values of L''_e, and these will again be special cases of cds. Note that every truth set of L''_e is a truth set of L'', but the converse is of course false.

3 Relations between the calculi L'' and L''_e

Let A be a sentence which involves q predicates $r_j^{p_j}$ of respective weights p_j. Consider the following sentences:

$$F_1 = \forall x\, [exx]$$

$$F_2 = \forall x\, \forall y\, \forall z\, [exy \to (exz \to eyz)]$$

$$G_{ji} = \forall y\, \forall x_1 \cdots \forall x_{p_j}\, [ex_iy \to (r_j^{p_j}\, x_1 \cdots x_{p_j} \to (y/x_i)\, r_j^{p_j}\, x_1 \cdots x_{p_j})]$$

$$\text{for} \quad 1 \leqslant j \leqslant q \quad \text{and} \quad 1 \leqslant i \leqslant p_j.$$

All these sentences are provable in L''_e (but not in L'').

Let $K(A)$ denote the (finite) set of these sentences.

1 Lemma *In the calculus L'': $K(A) \vdash \forall x\, \forall y\, [exy \to eyx]$ and*

$$K(A) \vdash \forall y\, \forall x_1 \cdots \forall x_{p_j}\, [ex_iy \to (r_j^{p_j}\, x_1 \cdots x_{p_j} \leftrightarrow (y/x_i)\, r_j^{p_j} x_1 \cdots x_{p_j})].$$

Proof $K(A) \vdash F_2$, and hence $K(A) \vdash eab \to (eac \to ebc)$ for any a, b, c. Thus $K(A), eab, eac \vdash ebc$. In particular, for $c = a$ we get $K(A), eab, eaa \vdash eba$. But since $K(A) \vdash F_1$, it follows that $K(A) \vdash eaa$, and so $K(A), eab \vdash eba$, that is, $K(A) \vdash eab \to eba$. By the principle of generalization we finally obtain $K(A) \vdash \forall x\, \forall y\, [exy \to eyx]$.

For the second assertion of the lemma: $K(A) \vdash G_{ij}$, so that

$$K(A) \vdash ea_ib \to (r_j^{p_j} a_1 \cdots a_{p_j} \to (b/a_i)\, r_j^{p_j} a_1 \cdots a_{p_j}),$$

for any choice of the individuals $a_1, \ldots, a_{p_j}$—we assume them distinct.

Thus, by interchanging a_i and b, we have

$$K(A) \vdash eba_i \to ((b/a_i)\, r_j^{p_j} a_1 \cdots a_{p_j} \to (a_i/b)\, (b/a_i)\, r_j^{p_j} a_1 \cdots a_{p_j}).$$

Thus

$$K(A) \vdash eba_i \to ((b/a_i)\, r_j^{p_j} a_1 \cdots a_{p_j} \to r_j^{p_j} a_1 \cdots a_{p_j}).$$

But

$$K(A), ea_ib \vdash K(A), eba_i,$$

and thus

$$K(A) \vdash ea_ib \to ((b/a_i)\, r_j^{p_j} a_1 \cdots a_{p_j} \to r_j^{p_j} a_1 \cdots a_{p_j}).$$

It follows that $K(A) \vdash ea_ib \to ((b/a_i)\, r_j^{p_j} a_1 \cdots a_{p_j} \to r_j^{p_j} a_1 \cdots a_{p_j})$. The assertion now follows by the principle of generalization.

2 Lemma *Let B be a sentence containing none of the predicates figuring in A. Then (in L''): $K(A) \vdash eab \to (B \to (b/a)\, B)$ for any a and b.*

Proof Let h be a system of truth values of L'' such that

$$\tilde{h}(F_1) = \tilde{h}(F_2) = \tilde{h}(G_{ij}) = 1 \text{ for all } i \text{ and } j;$$

$$\tilde{h}\,(eab) = 1.$$

We shall show that $\tilde{h}(B) = \tilde{h}\,((b/a)\,B)$, which will complete the proof (by the completeness of the calculus L'').

We reason by induction on the order of B.

If $B \in \mathfrak{a}$, the assertion is an immediate consequence of Lemma 1.

The rest of the proof proceeds as in the proof of the Consistency Theorem for L''_e (see p. 120).

Using these two lemmata, we can prove

1 THEOREM *$A \in T''$ (in L''_e) if and only if $K(A) \vdash A$ (in L'').*

Proof Suppose that $K(A) \vdash A$. Then $\wedge K(A) \to A \in T$. But $T \subset T_e$, and so $\wedge K(A) \to A \in T_e$. Hence $K(A) \vdash_e A$; but $K(A) \subset T_e$, and so $A \in T_e$.

Now suppose that $A \in T_e$. We know (Chapter I, Proposition 9) that A has a formal proof $A_1, \ldots, A_n$ whose terms contain only predicates occurring in A. We shall show by induction on i that $K(A) \vdash A_i$.

A_1 is an instance of an axiom schema of L''_e; if it is one of S1, ..., S11, it is also an instance of an axiom schema of L''_e, so that $K(A) \vdash A_1$; if $A_1 = eaa$ or $A_1 = eab \to (B \to (b/a)\,B)$, then $K(A) \vdash A_1$ by the definition of $K(A)$ and Lemma 2.

If A_i follows from sentences A_j and $A_h = A_j \to A_i$ by modus ponens, and we assume that $K(A) \vdash A_j$ and $K(A) \vdash A_j \to A_i$, it is clear that $K(A) \vdash A_i$.

If $A_i = \exists x\,[f^x] \to B$ is obtained from $A_j = (a/x) f^x \to B$, where $a \notin I_{B,f^x}$ and $K(A) \vdash (a/x) f^x \to B$, then $K(A), (a/x) f^x \vdash B$. Since $K(A)$ does not contain any individuals, it follows that $K(A), \exists x\,[f^x] \vdash B$, whence again $K(A) \vdash A_i$.

It now follows immediately that $K(A) \vdash A$.

4 Completeness theorem

We now propose to show that any consistent sentence A is satisfiable in L''_e. This will imply the Semantic Completeness Theorem; for if A is universally valid in L''_e, then $\neg A$ is not satisfiable, therefore $\neg A$ is inconsistent and so $A \in T_e$.

Let A be consistent in L''_e: $\neg A \notin T_e$. Then, by the above theorem, $\neg A \notin T\,(K(\neg A))$. But $K(\neg A) = K(A)$, and therefore $\neg A \notin T\,(K(A))$. Thus $K(A), A$ is consistent in L''.

Since the set $K(A)$, A is finite, it follows that there exists a truth set V_h in L'' such that $K(A), A \subset V_h$ (Completeness Theorem for L'').

Let $\mathscr{B}$ be the set consisting of all atomic equations and all atomic sentences of the form $r_n^p a_1 \cdots a_p$, where r_n^p is a predicate figuring in A (and $a_1, \ldots, a_p$ are any individuals). Define a system of truth values h' by

$$h'(u) = \begin{cases} h(u) & \text{if} \quad u \in \mathscr{B}, \\ 0 & \text{if} \quad u \notin \mathscr{B}. \end{cases}$$

We claim that h' is a system of truth values for L_e''.

Suppose that $h'(eab) = 1$, and let u be any atomic sentence:

If $u \notin \mathscr{B}$, then $h'(u) = h'((a/b)\, u) = 0$.

If $u = ecd$, where $b \neq c$ and $b \neq d$, then $u = (a/b)\, u$.

Let $u = ebd$ where $b \neq d$.

$$K(A) \vdash eab \to (ead \leftrightarrow ebd),$$

so that $h(eab) \to (h(ead) \leftrightarrow h(ebd)) = 1$. But $h(eab) = 1$, so that

$$h(ead) = h(ebd),$$

thus

$$h(u) = h((a/b)\, u)$$

or

$$h'(u) = h'((a/b)\, u).$$

If $u = ecb$ or $u = ebb$, the proof is analogous.

If $u = r_n^p a_1 \cdots a_p$, the proof is similar (cf. Lemma 2).

Conversely, if $h'(u) = h'((a/b)\, u)$ for all $u \in \mathfrak{a}$, then, in particular, for $u = eab$:

$$h'(eab) = h'(eaa) = h(eaa) = 1, \text{ since } K(A) \vdash eaa.$$

Thus h' is indeed a system of truth values for L_e'', which coincides with h for atomic subsentences of A. Thus $\tilde{h}'(A) = \tilde{h}(A) = 1$, and it follows that A is satisfiable in L_e''.

Generalizations

Let J be any set of sentences, and define $\hat{J} = J \cup \bigcup_{A \in J} K(A)$. Then:

2 THEOREM *J is consistent in L_e'' if and only if $\hat{J}$ is consistent in L''.*

Proof If $\hat{J}$ is inconsistent in L'', it has a finite subset $\hat{J}_0$ which is inconsistent. Thus $\hat{J}_0 \vdash \omega$, i.e. $\wedge \hat{J}_0 \to \omega \in T$. But $T \subset T_e$, and hence $\hat{J}_0 \vdash_e \omega$, and so $\hat{J} \vdash_e \omega$. Now $\bigcup_{A \in J} K(A) \subset T_e$, and it follows that $J \vdash_e \omega$, so that J is inconsistent in L_e''.

Conversely, if J is inconsistent in L_e'', it has a finite subset J_0 such that $J_0 \vdash_e \omega$, or $\wedge J_0 \rightarrow \omega \in T_e$. Hence $K(\wedge J_0 \rightarrow \omega) \vdash \wedge J_0 \rightarrow \omega$. It follows that $\hat{J}_0 \vdash \omega$, and J is inconsistent in L''.

3 THEOREM *If J is a consistent and denumerable subset (in L_e''), then J is satisfiable in a denumerable subdomain I^* of I.*

Proof J is also consistent and denumerable in L'', thus satisfiable in a denumerable subdomain I^*. In other words, there exists a mapping $s: I \hookrightarrow I^*$, which is one-to-one on I_J, such that $S(J)$ is contained in a truth set V_h^* of E^*.

Thus it suffices to define $h': \mathfrak{a}^* \hookrightarrow \mathbb{U}$ by setting $h'(u) = h(u)$ when u is an atomic equation or an atomic sentence $r_n^p a_1 \cdots a_p$, where r_n^p is a predicate contained in at least one sentence of J, $h'(u) = 0$ otherwise.

One then verifies, as in the preceding pages, that h' is a system of truth values for L_e''. It then follows that $s(J) \subset V_h^*$.

This result may also be generalized for sets of sentences of arbitrary cardinality.

References

ARENS, R.F. and KAPLANSKY, I., Topological representations of algebras, *Trans. Amer. Math. Soc.*, **63**, 457–481 (1948).

BETH, E.W., A topological proof of the theorem of Löwenheim-Skolem-Gödel, *Indagat. Math.* **13**, 436–444 (1951).

BETH, E.W., *The Foundations of Mathematics*, North-Holland Publishing Company, Amsterdam (1959).

CHURCH, A., *Introduction to Mathematical Logic*, Vol. 1, Princeton University Press, Princeton, N.J. (1956).

GENTZEN, G., *Recherches sur la Déduction Logique*, Ladrière, Paris (1955).

HALMOS, P.R., *Algebraic Logic*, Chelsea, New York (1962).

HALMOS, P.R., *Lectures on Boolean Algebras*, Van Nostrand, Princeton (1963).

HILBERT, D. and ACKERMANN, W., *Mathematical Logic*, Chelsea Publishing Company, New York (1950).

KLEENE, S.C., *Introduction to Metamathematics*, North-Holland Publishing Company, Amsterdam (1952).

MENDELSON, E., *Introduction to Mathematical Logic*, Van Nostrand, London (1965).

PONASSE, D., Problèmes d'universalité s'introduisant dans l'algébrisation de la logique mathématique, *Nagoya Math. J.*, **20**, 29–73 (1962), **21**, 61–110 (1962).

QUINE, W.V.O., *Methods of Logic*, Henry Holt & Company, New York (1959).

RASIOWA, H. and SIKORSKI, R., A proof of the completeness theorem of Gödel, *Fund. Math.* **37**, 193–200 (1950).

RASIOWA, H. and SIKORSKI, R., A proof of the Skolem-Löwenheim Theorem, *Fund. Math.* **38**, 230–232 (1951).

SIKORSKI, R.. *Boolean Algebras*, Springer-Verlag, Berlin (1964).

STONE, M.H., The theory of representations for Boolean algebras, *Trans. Amer. Math. Soc.*, **40**, 37–111 (1936).

STONE, M.H., Applications of the theory of Boolean rings to general topology, *Trans. Amer. Math. Soc.*, **41**, 321–364 (1937).

List of symbols